Optical Measurements for Scientists and Engineers

A Practical Guide

With this accessible, introductory guide, you will quickly learn how to use and apply optical spectroscopy and optical microscopy techniques. Focusing on day-to-day implementation and offering practical lab tips throughout, it provides step-by-step instructions on how to select the best technique for a particular application, how to set up and customize new optical systems, and how to analyze optical data. You will gain an intuitive understanding of the full range of standard optical techniques, from fluorescence and Raman spectroscopy to super resolution microscopy, and understand how to navigate around an optics lab, with clear descriptions given of the most common optical components and tools. Including explanations of basic optics and photonics, and easy-to-understand mathematics, this is an invaluable resource for graduate students, instructors, researchers, and professionals who use or teach optical measurements in laboratories.

Arthur McClelland is a Principal Scientist at the Center for Nanoscale Systems at Harvard University. He is a member of the American Chemical Society and the New England Society for Microscopy, and has co-taught Harvard Extension School's *Introduction to Microscopy* course.

Max Mankin is the Co-Founder and CTO of Modern Electron, USA, and completed his PhD at Harvard University.

Spectroscopy and Microscopy are two important fields in today's technology. *Optical Measurements for Scientists and Engineers* provides an excellent introduction to these fields with a very practical point of view. The book is very well illustrated and explained; and it limits the number of equations used. Thus *Optical Measurements for Scientists and Engineers* is within the reach of a wide audience. Of course, the emphasis is on optical measurements, which are a critical aspect in applied technology.

José Sasian, ***University of Arizona***

Optical Measurements for Scientists and Engineers

A Practical Guide

Arthur McClelland
Harvard University, Massachusetts

Max Mankin
Harvard University, Massachusetts

CAMBRIDGE
UNIVERSITY PRESS

CAMBRIDGE
UNIVERSITY PRESS

Shaftesbury Road, Cambridge CB2 8EA, United Kingdom

One Liberty Plaza, 20th Floor, New York, NY 10006, USA

477 Williamstown Road, Port Melbourne, VIC 3207, Australia

314–321, 3rd Floor, Plot 3, Splendor Forum, Jasola District Centre, New Delhi – 110025, India

103 Penang Road, #05–06/07, Visioncrest Commercial, Singapore 238467

Cambridge University Press is part of Cambridge University Press & Assessment, a department of the University of Cambridge.

We share the University's mission to contribute to society through the pursuit of education, learning and research at the highest international levels of excellence.

www.cambridge.org
Information on this title: www.cambridge.org/9781107173019

DOI: 10.1017/9781316779613

First published 2018

A catalogue record for this publication is available from the British Library

ISBN 978-1-107-17301-9 Hardback

For Simon, Tobi, and Lisa.
Thanks for being patient while I wrote a book.
A. M.

To my parents and my sister, who by example
taught me the value of hard work and the importance
of passion for one's pursuits.
M. M.

Contents

Preface

Experiments involving lasers, light, and optics are ubiquitous in virtually every field of experimental science and engineering. Recently, advanced optical characterization techniques have begun to pervade the life sciences and medical fields as more physical scientists are turning their attention to biological questions. There is still a strong "build it yourself" culture in optics and, unfortunately, graduate students and postdocs often find themselves suddenly in charge of a lab full of expensive optics that they are totally unfamiliar with. This book is meant as a crash course "What is that thing?"-type guide, complete with practical tips and tricks to help get optical setups built and collecting data. It focuses on the most common components used in optical spectroscopy and microscopy setups and it is by no means exhaustive. It is, of course, biased by the authors' own experiences in various academic optics labs but it is hoped that it will help the reader feel more comfortable in the lab, dispel some of the incorrect lab lore that is often passed down through the years, and result in better (and easier!) science at the end of the day.

Is this book right for you? We've written the text for biologists, chemists, materials scientists, applied physicists, electrical engineers, and medical professionals, and purposefully limited the theory and the math in favor of diagrams, pictures, tips, tricks, and quick, practical information that will help you get your measurement done without earning a second degree. To understand everything in this book, you'll only need a little algebra and trigonometry, an understanding of 45 and 90 degree angles, and maybe an introductory chemistry or physics course.

Perhaps you have encountered one of the following scenarios:

- You are asked to "Build a photoluminescence spectroscopy system" but you've never turned on a laser, let alone aligned a laser beam, mirrors, and lenses.
- Total internal reflectance microscopy sounds just perfect for characterizing your cells, but you don't know what such a microscope looks like or which vendors to buy the components from.
- You see a complicated graph in an exciting seminar and everyone is talking about it but you have no idea how to even begin interpreting all of the contour lines and symbols.

- You find an unlabeled, dirty filter on your laser table and the previous student (who has now graduated, moved 6000 miles away, and started a bakery) hinted some time last year that maybe it's the one that you need for your new measurement. How do you figure out what it is?
- Everyone seems to be speaking another language referring to lots of things in optics lab that just sound like gibberish to you. Can you please hand them the what?!

We, the authors, have experienced situations like these, so we wrote this book thinking about our experiences and state of mind when we first started working with laser tables and optics. A practical guide like this would certainly have saved us from many late nights of bad measurements and frustration with laser alignment. We hope it will be helpful to you and your science.

Acknowledgments

As with any major undertaking, there are many people whose contributions have led to this book, sometimes knowingly and sometimes unknowingly. I apologize in advance to anyone I inadvertently leave out. It has been an exciting time to be involved with spectroscopy and microscopy, and I look forward to what the future holds.

I started experimenting with lasers by making holograms and exploring optics in general at a summer science camp in Morgantown, WV, circa 1993. Thanks to my parents Christine and Steve McClelland for letting me enroll in the science summer camp. Thanks to Judy Werner for doing the thankless job of organizing the summer camp every year, and a special thanks to Sean Lalley for letting a bunch of middle schoolers play with lasers and develop holograms. Lasers in those days were expensive and often somewhat homebuilt. In retrospect, it was an amazing effort on his part to organize the hands-on making of holograms by kids.

In college I was fortunate enough to work in a number of research groups with quite a number of very patient graduate students and postdocs. Thanks to Sava Denev and Professor David Snoke for my first summer research experience in an optics lab at the University of Pittsburgh. Thanks to Vasiliy Fomenko for his infinite patience in teaching me how to realign a Ti:sapphire laser after I had messed it up yet again, and Professor Eric Borguet for giving me the opportunity to do an undergraduate research thesis in his lab in ultrafast optics at the University of Pittsburgh. Thanks to the Applied Physics program at the University of Michigan for support to explore different areas of research. Thanks to Professor Steve Yalisove for time in his group. Thanks to Jie Wang for the patience while teaching me how to realign an OPO/OPA/DFG system for SFG-VS spectroscopy. Thanks to Professor Zhan Chen for overseeing my PhD thesis at the University of Michigan. Thanks to Abdelkrim Benabbas and Professor Paul Champion for my postdoctoral experience at Northeastern University.

Thanks to Eric Martin and Fettah Kosar for hiring me and giving me a chance to grow professionally at the Center for Nanoscale Systems at Harvard University. Thanks especially to Fettah for getting the Introduction to Microscopy course for the Harvard Extension School off the ground, as large sections of this book grew out of lectures and discussions from this course. Thanks to David Bell, Bill Wilson, Greg Lin, and all my other CNS colleagues for their continued support.

Thanks to all the users of CNS who bring me new scientific challenges to think about every single day.

Thanks to Doug Richardson at the Harvard Center for Biological Imaging for contributing a variety of microscopy images and being a great colleague. I sometimes wish that all my colleagues were Canadian.

Thanks to Jennifer Waters in the Nikon Imaging Center at Harvard Medical School. The Quantitative Imaging course at Cold Spring Harbor Laboratories was a formative two-week experience. Thanks to the New England Society for Microscopy for providing a community of microscopists. Thanks to all the vendors who I have coerced into giving workshops at CNS over the years, especially Philippe Clemenceau for organizing the FLIM and New Approaches in Light Microscopy workshops with me. I learn a lot every time.

Thanks to Mac Hathaway and Andy Boughton for giving critical feedback on the manuscript. Thanks to Cambridge University Press for being patient while we worked on this book around our day jobs and family commitments.

Finally, thanks to my co-author Max Mankin who made the hypothetical idea of writing this book a reality with his infectious enthusiasm.

Arthur McClelland

When I started graduate school, I had never worked in an optics lab before. There were lots of people who helped me learn my way around a laser table.

First, thanks to my co-author Arthur McClelland, who taught me optics, tolerated countless uninformed questions, and humored my suggestion to write a book to help others who wanted to use optical techniques without much of a background in it. Arthur served as a constant resource, inspiration, and friend in using optics for my research.

Second, I owe my PhD adviser Professor Charles M. Lieber a big debt of gratitude for figuratively throwing me in the deep end of the world of optics in one of my first graduate school projects. I learned by doing! Professor Lieber was also generous enough to provide me opportunities to learn from experts in this area, including Professor Ritesh Agarwal at the University of Pennsylvania and his accommodating and patient student Dr. Carlos Aspetti, and Professor Hongjie Dai at Stanford. During that trip to Stanford, I worked with Dr. Guosong Hong for the first time. Guosong patiently tolerated my rudimentary understanding of infrared optics and photonics and very quickly got me up to speed. I'm enormously thankful for the opportunities to learn from experts that I had during this time.

I was also lucky enough at the time to have fantastic mentors and experienced laser jocks to constantly pester with questions. Thank you to my fantastic colleagues Dr. Robert Day, Dr. Ruixuan Gao, Dr. Guosong Hong, Professor Thomas Kempa, Professor Bozhi Tian, Professor James Cahoon, Dr. Jinlin Huang, and Dr. Carl Barrelet, who helped me navigate my journey from zero to one around a laser table, doing everything from teaching me about how to build and automate measurement setups to identifying unlabeled filters and lenses.

I also owe special thanks to Professors Hong-Gyu Park, Sun-Kyung Kim, and You-Shin No, collaborators and true optics and photonics experts who patiently spent many coffee breaks and late evenings explaining the principles of optics and photonics to me.

I also owe an enormous thanks to my thesis committee members and mentors Professor David Bell and Professor Cynthia Friend, as well as the Harvard Center for Nanoscale Systems staff, including Adam Graham, Greg Lin, Fettah Kosar, Andrew Magyar, and Steve Hickman for their constant support and early introductions to many of the techniques listed in this book. Thanks as well to Professors Conor Evans and Ed Boyden, whose enthusiasm and deep understanding of optics in biology was educational and inspirational. I also owe early inspiration and education in the principles of physical chemistry and optics to Brown Professors Shouheng Sun, Jimmie Doll, Eunsuk Kim, my mentor Dr. Vismadeb Mazumder, as well Professors Krysta Ryzewski, Brian Sheldon, and Susan Alcock, who served as inspirational mentors in testing cultural heritage artifacts using optical and other analytical techniques.

Finally, thank you to all of my friends, colleagues, and family who encouraged me to write this book! Hearing from you about your experiences in optics and photonics labs informed the contents, writing style, and level of detail that we hope will make this a useful manual for everyone out there who needs to use optical characterization for their science, engineering, and analysis.

Max Mankin

1 Introduction

Science experiments are performed to gain information about a sample. There are many different ways to interrogate a sample. Nature is exacting though, and will only answer the specific question that is asked. The different techniques discussed in this book are all asking slightly different questions, usually with light. It is important to keep in mind exactly what question you are asking with each experimental technique in order to correctly interpret the data.

Light carries information. This is a critical point to keep in mind. For instance, light reflecting off of objects brings information to your eyes and brain about your surroundings. Light from the sun provides astronomers information about our nearest star's composition and temperature. And light transmitted through undersea cables enables sharing cat videos and stock market fluctuations with your friends and colleagues on other continents.

Experimental optics is all about using light to bring you the information you need about your sample and the physical world. Naturally, light can carry many different types of information. As you use optics to ask physical questions of your samples, it is essential that you can control the light you are using and choose the correct form of it so that you can extract the right type of information.

This book aims to help you do exactly that. We cover the basic principles of light and optics in Chapter 1, and then the different optical components used to manipulate light in Chapter 2. Chapter 3 is about spectroscopic techniques, which provide a wealth of information about the sample. Chapter 4 is about the plethora of optical microscopy techniques. Finally, Chapter 5 covers some of the practical tips and tricks for setting up your own optical experiments.

If you take nothing else from this book, remember that:

- light carries information;
- to get the information you want, you need to choose the correct form of light.

1.1 Light: A Brief Introduction to its Properties

This section is a quick introduction to some important concepts about light that will be used in the rest of the book. It is by no means an exhaustive course in optics.

1.1.2 What is Light?

Light is electromagnetic radiation. It can be modeled in two ways: as a wave or as a particle (a "photon"). Scientists switch back and forth between the two models of light depending on the physical phenomenon that they are describing (Appendix 2).

The wave-like nature of light was demonstrated by Thomas Young's double slit experiment in 1801 in which he observed constructive and destructive interference between two light beams (Appendix 3). The particle nature of light was used by Albert Einstein to explain the photoelectric effect (Appendix 1), for which he was awarded the Nobel Prize in 1921. In the photoelectric effect, light incident on a metal ejects electrons. The more intense the light, the more electrons are ejected, but all the electrons have the same energy. This means that each photon of the same color has the exact same amount of energy. If light acted only as a wave, illuminating the sample with light of higher intensity should produce ejected electrons with higher energies.

Which model of light is correct? Both and neither are correct. By using the appropriate model, different optical phenomena can be quite accurately described, but clearly a perfect model would describe all optical phenomena without the need to jump back and forth between two different models. The mathematically unified description of electromagnetic radiation is beyond the scope of this book, so for now please accept the need to jump back and forth between conceptual models.

In the wave model, light consists of oscillating electric and magnetic fields that are both orthogonal (i.e., perpendicular) to each other and orthogonal to the propagation direction of the light (Figure 1.1). The electric field of light is the component that predominantly interacts with matter (e.g., your sample), so to describe light–matter interactions you can usually ignore the magnetic field. The intensity of the light is the magnitude of the electric field squared.

Light waves can be described by their wavelength or their frequency. Wavelength is the distance from one peak in the wave to the next peak in the wave. Wavelength is a distance and is measured in meters. Visible light is usually described in terms of wavelength. Visible light for humans ranges from 400 to 800

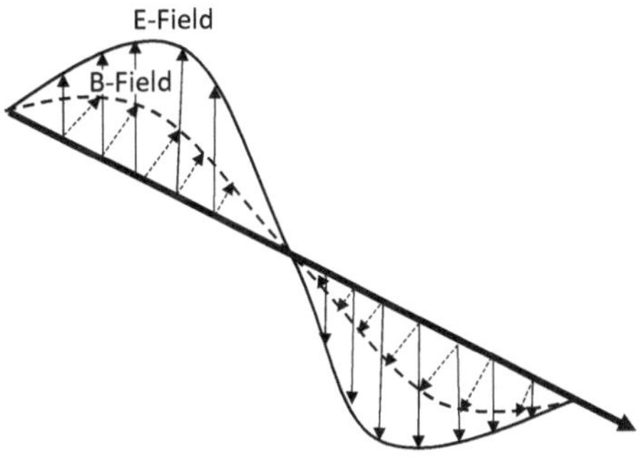

Figure 1.1

Light is electromagnetic radiation that consists of perpendicular electric and magnetic fields that are orthogonal to the direction of the propagation of light. The electric field is traditionally denoted with E and the magnetic field is traditionally denoted with B. The magnetic field is largely ignored in optics and only the interaction of the electric field with matter is considered.

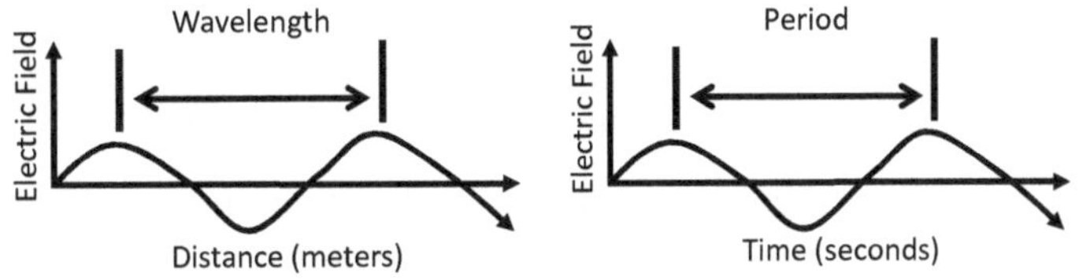

Figure 1.2

Waves can be described in terms of either wavelength or frequency, which equals 1/period.

nanometers (nm) in wavelength. One nanometer is 10^{-9} meters. Frequency is the reciprocal of the period (the time for the electric field to go from one peak to the next). Frequency is measured in hertz (Hz), which describes the number of peaks, or cycles, per second. Radio waves, also an electromagnetic wave, are usually described by their frequency. The local radio station is 90.9 MHz, for example.

In the particle model, light can be described by the energy of a single photon. The energy of a photon is usually described in units of electron volts (eV). Photons have no mass, so they can be easily created and destroyed without violating the law of conservation of mass. Photons do obey the law of conservation of energy. Also, despite having no mass, photons do have momentum, so they do obey conservation of momentum too. This property can be taken advantage of for solar sails for interplanetary probes and for optical tweezers and optical traps in a microscope.

In the particle model of light, each photon contains a discrete, well-defined, amount of energy. The energy of a photon is directly related to the frequency in the wave model and thus wavelength by the de Broglie relation:

$$E = h\nu = \frac{hc}{\lambda} \tag{1.1}$$

where E is the energy of a single photon;

h is Planck's constant (6.626×10^{-34} Js, where J stands for Joules, a unit of energy, and s stands for seconds);
ν is the frequency of light; and
λ is the wavelength of light.

TIP: The shorthand "hν" is often used instead of the word "photon" in pictures and diagrams.

There is a direct relation between the wavelength and the frequency of light:

$$c = \lambda\nu \tag{1.2}$$

where c is the speed of light in a vacuum (3×10^8 m/s);

λ ("lamb-dah") is the wavelength of light; and
ν ("new") is the frequency of light.

From Equation (1.2), a 400 nm photon (a purple photon) contains twice the energy of an 800 nm photon (a red photon). The perceived color of light is thus directly related to the energy of the photon or the wavelength and frequency of the light wave. Scientists often refer to things as "red shifted" when the light shifts down in energy and "blue shifted" when light shifts up in energy.

The electromagnetic spectrum is traditionally split into different ranges. The visible light range (400–800 nm, spanning violet to red) is the one that we are most familiar with. Just outside the visible light range are ultraviolet (UV) light (100–400 nm) range and infrared (IR) light (800–50,000 nm) range. The IR region is commonly split into the near IR (800–3,000 nm), mid-IR (3–25 μm), and far-IR regions (25–50 μm). Light below 190 nm is referred to as vacuum UV (VUV) because it is strongly absorbed by molecules in the air, and can only propagate through a vacuum. Light in the 1–100 nm wavelengths are referred to as extreme UV (EUV). Light with a wavelength shorter than the UV range is commonly referred to as soft x-rays, then hard x-rays, then gamma rays. Light with wavelengths longer than that in the IR region is referred to as microwaves and then radio waves. The exact cutoffs for the different regions vary depending on who you are talking to, but the values given in Table 1.1 are good general guides.

Table 1.1 Different regimes of the electromagnetic spectrum.

	Wavelength	Energy	Frequency	Example
Gamma rays	<0.01 nm	>124 keV	>3×10^{13} MHz	Radioactive decay
Hard x-rays	0.1–0.01 nm	12.4–124 keV	3×10^{12}–3×10^{13} MHz	Synchrotron
Soft x-rays	0.1–10 nm	124 eV–12.4 keV	3×10^{10}–3×10^{12} MHz	
Extreme ultraviolet (EUV)	1–100 nm	12.4 eV–1.24 keV	3×10^{9}–3×10^{11} MHz	Lithography (Section 4.20)
Vacuum ultraviolet (VUV)	100–190 nm	6.53–12.4 eV	3×10^{9}–1.5×10^{9} MHz	
Ultraviolet light (UV)	190–400 nm	3.1–6.53 eV	1.5×10^{9} MHz–7.5×10^{8} MHz	Sun burn
Visible light (VIS)	400–800 nm	1.55–3.1 eV	7.5×10^{8}–3.7×108 MHz	What you see
Near infrared (NIR)	800–3000 nm	0.413–1.55 eV	3.7×10^{8}–1.0×108 MHz	Night vision goggles
Mid-infrared (MIR)	3–25 μm	50–413 meV	1.0×10^{8}–1.2×10^{7} MHz	Heat/thermal imaging
Far-infrared (FIR)	25–50 μm	25–50 meV	1.2×10^{7}–5.88×10^{6} MHz	Rarely used
Terahertz (THz)	10 μm–1 mm	124–1.24 meV	0.3–30 THz	Medical imaging, security
Microwaves	1 mm–1 m	1.24 meV–1.24 μeV	300 MHz–300 GHz	Wi-Fi, cell phone communications
Radio waves	>1 m	<1.24 μeV	<300 MHz	AM/FM radio

1.2 An Explanation of Energy, Wavelength, and Frequency Jargon

Note that various frequency, wavelength, and energy units are preferred for different wavelength regimes, since using so many powers of 10 with a single unit is quite impractical. Table 1.2 shows the preferred units for each regime of the electromagnetic spectrum. Note that the unit cm^{-1} is known as the "wavenumber" and is a unit of energy corresponding to 0.00012 eV that is typically used to describe the MIR and FIR infrared regimes. For instance, one might say, "a peak

Table 1.2 The most commonly used units for various regions of the electromagnetic spectrum

	Wavelength	Energy	Frequency
Gamma rays	–	keV, MeV, GeV	–
Hard x-rays	Å, nm	keV	–
Soft x-rays	Å, nm	eV, keV	–
Extreme ultraviolet (EUV)	nm	eV, keV	–
Vacuum ultraviolet (VUV)	nm	eV	–
Ultraviolet light (UV)	nm	eV	–
Visible light (VIS)	nm	eV	–
Near infrared (NIR)	nm, μm	eV, meV	cm^{-1}
Mid-infrared (MIR)	μm	–	cm^{-1}
Far infrared (FIR)	μm	–	cm^{-1}
Terahertz (THz)	μm, mm	–	THz
Microwaves	mm, m	–	MHz, GHz
Radio waves	cm, m, km	–	kHz, MHz

at 1730 wavenumbers corresponds to a C=O stretch in infrared spectroscopy" (Section 3.4 for more on infrared spectroscopy).

As shown in Table 1.3, different energy regimes are associated with different physical phenomena in atoms, molecules, and solids. Therefore, one can use different optical techniques associated with different energy regimes to ask different questions of a sample.

TIP: Note that Naomi Halas' group at Rice University hosts a truly fantastic tool to convert between the various energy (eV), frequency (cm^{-1}, Hz, GHz, THz), and wavelength (nm, μm) units. As of press time for this book, it can be found at: http://halas.rice.edu/conversions.

1.3 Polarization

Polarization refers to the direction of the electric field. Often, the polarization of light has an effect on the interaction with the sample, and some attention should be paid to the polarization of the light when setting up a new optical system.

In unpolarized light (also known as randomly polarized light), the electric field does not have a specific orientation with respect to the laboratory. Unpolarized

Table 1.3 Energy regimes for physical phenomena in atoms, molecules, and solids.

Physical phenomenon	Typical energy regime and units	Technique (section)
Molecular rotation and torsion 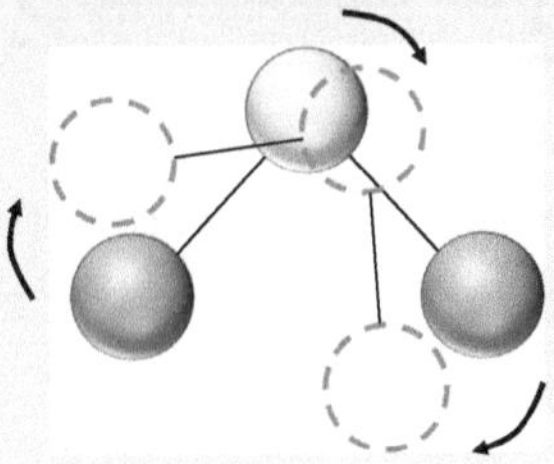	< ~250 GHz	Microwave spectroscopy (Not covered in this book)
Molecular and lattice vibration and rotation 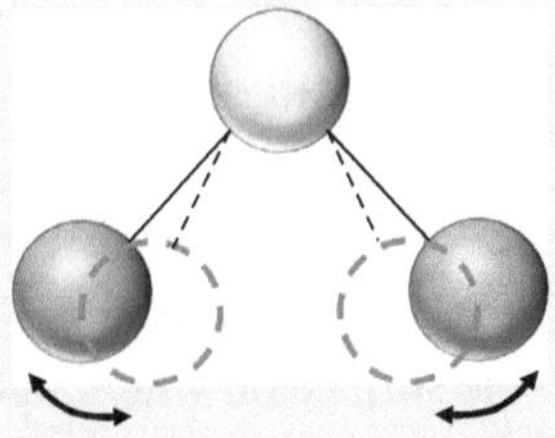	10–5000 cm^{-1} (~0.001–1 eV)	(Fourier transform) infrared spectroscopy (3.4) Raman spectroscopy (3.5)
Electronic transitions 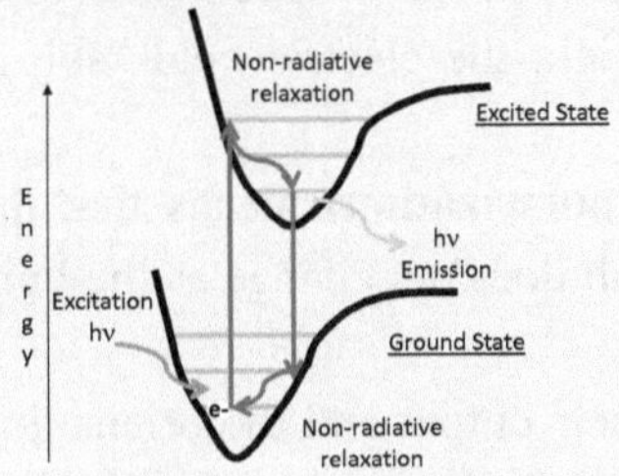	1–10 eV (150–1240 nm)	UV-VIS-NIR spectroscopy (3.1) Fluorescence spectroscopy (3.2) Photoluminescence (3.3) Ultrafast spectroscopy (3.8)
Core electron excitation and ejection 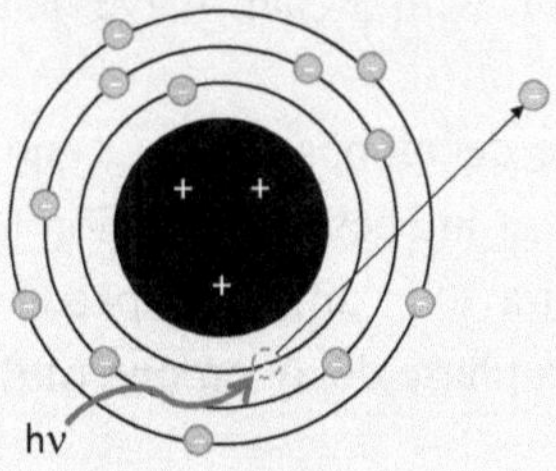	>10 eV	Photoelectron spectroscopy (Appendix 1)

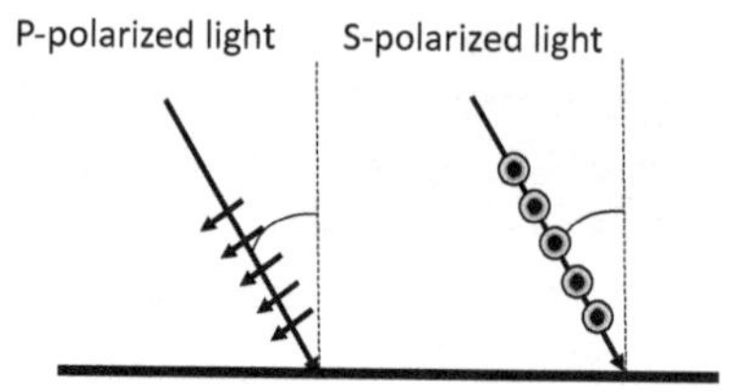

Figure 1.3

Polarization of light refers to the direction of the electric field. In P-polarized light the electric field is in the plane of the page and "plunges" into the surface. In S-polarized light the electric field is in and out of the page and "skips" on the surface.

light can be made linearly polarized with a polarizer (Section 2.3.14). Unpolarized light can also become partially polarized as it travels through a series of mirrors. In experiments where polarization is important, a "scrambler" (Section 2.3.16) is placed in the beam to restore a truly random polarization. Some optical components such as diffraction gratings in spectrometers have different throughput efficiencies for different polarizations of light.

Scientists describe the polarization of light by the letters S and P for historic reasons, using abbreviations from German terms for parallel and perpendicular. In P-polarized light, the polarization is perpendicular to the surface on which light is incident (in the plane of the page in Figure 1.3). In S-polarized light, the polarization is parallel to the surface (in and out of the page in Figure 1.3). A commonly used mnemonic is that in P-polarized light the electric field "plunges" into the surface, while in S-polarized light the electric field "skips" off the interface.

Linear polarization (sometimes called **plane polarization**) means that the orientation of the electric field with respect to the lab does not change as the light propagates (Figure 1.4).

Lasers are almost always linearly polarized at their output and the orientation of the electric field is usually set up to be either parallel to the plane of the optical table or perpendicular to the plane of the optical table. The polarization of the light can be flipped between S and P polarizations with mirrors arranged in a folded periscope configuration (Section 5.5) or with a half-wave plate (Section 2.3.15).

The electric field of the light is a vector that can be decomposed into x and y components. An **anisotropic** material can have different indices of refraction for the x direction and the y direction, meaning light will travel at different speeds in the two directions of the material. This will introduce a phase delay on one of the vector components of the electric field.

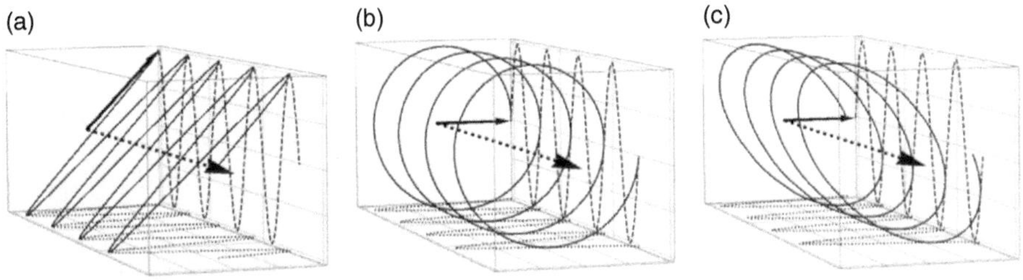

Figure 1.4

The solid arrows denote the direction of the electric field. The solid lines show how the direction of the electric field changes as the light propagates. The dotted arrows show the direction the light is traveling. The dotted and dashed lines depict how the *x* and *y* vector components of the electric field vary as the light travels. (a) Linearly polarized light means that the electric field direction does not change orientation with respect to the lab as the light propagates. The vector components of the electric field are in phase. (b) Circularly polarized light means the direction of the electric field rotates as the light propagates past a fixed point in the lab. The vector components of the electric field are out of phase by $\pi/2$ for circularly polarized light. (c) In elliptically polarized light, the orientation of the electric field spends more time in one direction than another as it rotates. The vector components of the electric field are out of phase for elliptically polarized light. The degree of the phase difference changes the ellipticity of the electric field.

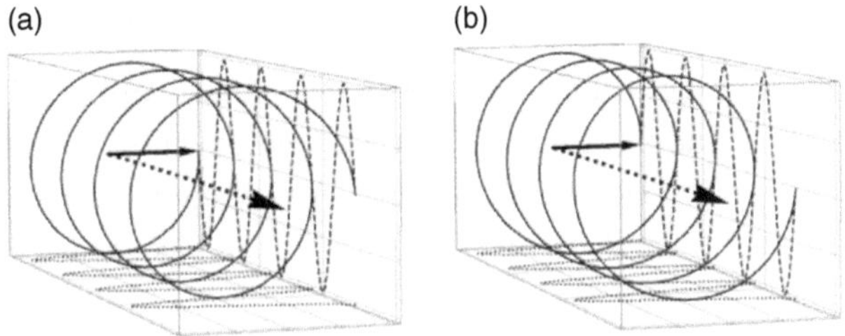

Figure 1.5

The solid arrows denote the direction of the electric field. The solid line shows how the direction of the electric field changes as the light propagates. The dotted arrow shows the direction the light is traveling. The dotted and dashed lines depict how the *x* and *y* vector components of the electric field vary as the light travels. The relative phase delay between the vector components can determine if the electric field will rotate clockwise or counterclockwise as the light travels. (a) In right-handed circularly polarized light, the electric field rotates clockwise. (b) In left-handed circularly polarized light, the electric field rotates counterclockwise.

If one direction is retarded by exactly $\pi/2$ relative to the other vector component, the linearly polarized light will become circularly polarized. In **circularly polarized light**, the electric field rotates around the axis of light propagation as the light travels. This means that in a given instant, the light incident on your sample will have an electric field oriented in a specific direction. But, averaged over a

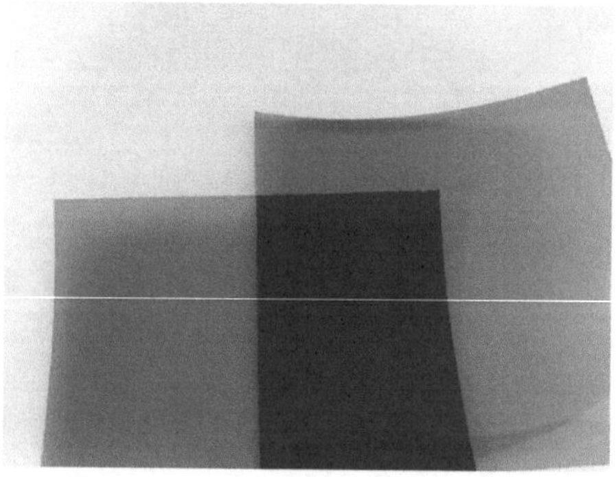

Figure 1.6

Two partially overlaid sheet polarizers. When the orientations of the polarizers are 90 degrees relative to each other, they are said to be crossed. Crossed polarizers will block all light as seen in the dark overlapping region in the middle. The single polarizer looks gray as it is blocking light of a specific polarization (about half the light in a typical unpolarized beam).

short amount of time, the sample will experience all directions of the electric field in turn as the light propagates.

If one vector component is delayed by some other amount, the light will become elliptically polarized. **Elliptical polarization** is similar to circular polarization, in that the electric field direction rotates around the axis of light propagation. However, the x and y vector components of the electric field are out of phase by a value other than $\pi/2$, so the electric field traces out an ellipse (Figure 1.4) instead of a circle.

Details of how to manipulate and control the polarization of light are given in Sections 2.3.14, 2.3.15, 2.3.16, and 5.5.

1.4 Spatial Resolution

"What is the smallest object that can be seen?" is a common first question for any imaging system. The answer is that it depends on the spatial resolution of the system. Spatial resolution is how close two objects can be and still be definitively observed as two objects. There are several mathematical definitions of spatial resolution. A commonly used definition is known as the Rayleigh criterion.

A diffraction-limited spot does not truly focus to a single point. It in fact has a series of rings around it referred to as Airy disks. The Rayleigh criterion says two points are just resolvable if the maximum of one Airy disk is at the first minimum of the other point's Airy disk.

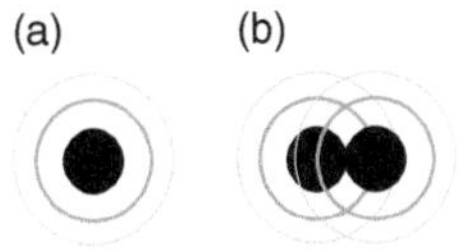

Figure 1.7

(a) Cartoon representation of an Airy disk formed by diffraction limited focused spot. (b) Two Airy disks at the Rayleigh criterion limit of resolvability.

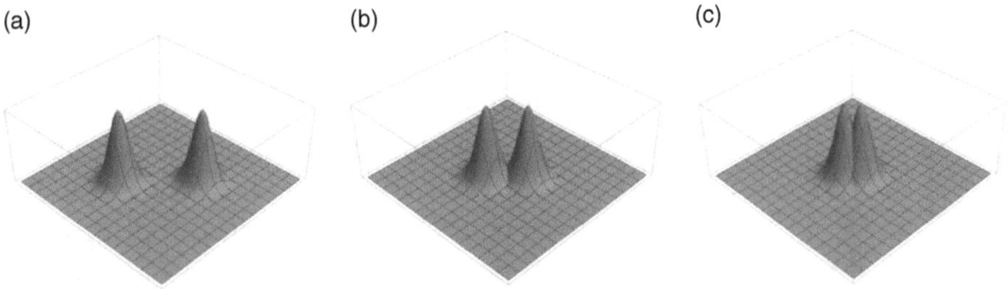

Figure 1.8

(a) Two fully resolved points. (b) Two overlapping but resolvable points. (c) Two unresolvable points.

Light with shorter wavelengths can probe smaller features. Resolution limits can be expressed quantitatively as a **diffraction limit**, *d*, which describes the smallest feature that it is physically possible to observe using light of a given wavelength, λ:

$$d = \frac{\lambda}{2n\, sin\, \theta} \sim \frac{\lambda}{2} \tag{1.3}$$

This limit was first described by Ernst Abbe in 1873. For instance, visible light (wavelength ~500 nm) has a diffraction limit ~250 nm. Therefore, it is possible to resolve features such as bacteria, cells, hair, and 1980s transistors (~1–100 μm), but not modern transistors, viruses, or cell walls (7–200 nm). To achieve higher resolution, one can use smaller wavelengths of light for imaging (e.g., UV or x-ray).

There is an important distinction between detection limit and resolution limit. Objects smaller than the resolution limit can be detected, but their spatial position cannot be localized less than the diffraction limit. For instance, a single fluorescent molecule (1–2 nm in length) can be detected, but the position cannot be determined with far-field optical means to better than ~200 nm. Several super resolution microscopy techniques have been developed to get around this limit. They are discussed in detail in Sections 4.13–4.17.

1.5 Near-Field and Far-Field

The rules surrounding diffraction limit and resolution apply in the case known as **far-field**, which is that the optical element is >~1–2 wavelength distances away from the object of interest. One can exceed the diffraction limit using **near-field** techniques, in which an object is effectively ~<1–2 wavelengths away from an optical element. Near-field techniques often involve a scanning probe approach. For example, the probe can be a tapered optical fiber that is metal coated and has a sub-diffraction limit aperture at the tip to allow an evanescent electric field of the light to leak out. Since the aperture is smaller than the diffraction limit, the light cannot propagate through the hole. With this approach, you have a localized electric field from the light that you can scan around the sample and test the sample localized electric field interactions.

The coupling of light into these scanning near-field optical microscope (SNOM) tips has proved hard enough that a new approach of scattering the light off the outside of a metal-coated atomic force microscopy tip and using a lock-in amplifier to separate the near-field and far-field is gaining popularity. This technique is referred to as scattering scanning near-field optical microscopy (S-SNOM), and is briefly described in Section 3.4.15.

1.6 Light–Matter Interactions

When light encounters solids, liquids, and gases, it can travel through (**transmission**), bounce off (**reflection** and **scattering**), or be taken up by a material (**absorption**).

We describe transmission and reflection by a quantity known as the refractive index, often written as n. Refractive index can be thought of qualitatively as how fast light can propagate through a material relative to the speed of propagation through vacuum. For instance, air has a refractive index of ~1 whereas water has a refractive index of ~1.33. This means that light travels 33 percent slower in water than in air. Glass has an index of refraction of around 1.5, which puts a limit on the speed of fiber optic communication.

Whenever light encounters an interface between materials with two different indices of refraction, some light will be reflected (bounce back) and some will be refracted (travel into the material but with a bend in the direction of propagation). The angle at which the light will be refracted is described by "Snell's law":

$$n_1 \sin \theta_1 = n_2 \sin \theta_2 \tag{1.4}$$

where n_1 and n_2 are the refractive indices of material 1 and material 2, respectively, and θ_1 and θ_2 are angles from the surface normal describing the direction from which the light propagates (Figure 1.9).

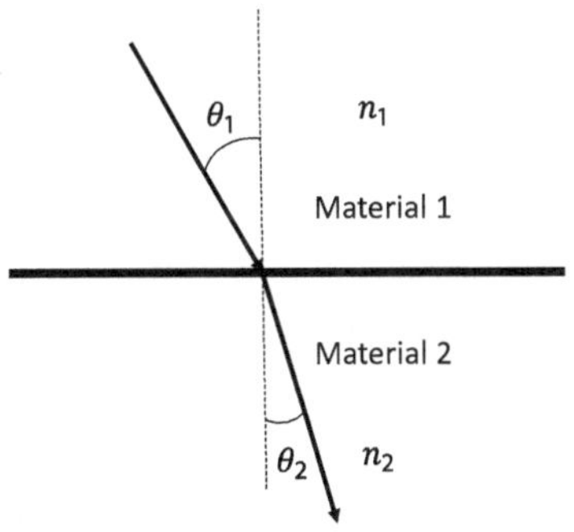

Figure 1.9

Snell's law describes how the direction of propagation of light will change when it encounters a material with a different index of refraction.

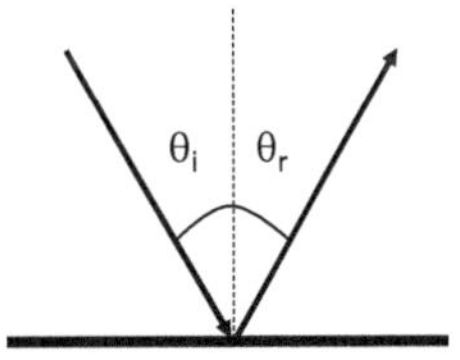

Figure 1.10

In specular reflection the angle of incidence θ_i is equal to the angle of reflection θ_r.

1.7 Reflection

When light reflects off a surface the angle of incidence (θ_i) is equal to the angle of reflection (θ_r).

In a dielectric (non-electrically conductive) material (e.g., a piece of glass) the amount of light that is reflected or transmitted is determined by the Fresnel equations:

$$R_p = |r_p|^2 = \left|\frac{\tan(\theta_i - \theta_t)}{\tan(\theta_i + \theta_t)}\right|^2 \tag{1.5}$$

$$R_s = |r_s|^2 = \left|\frac{\sin(\theta_i - \theta_t)}{\sin(\theta_i + \theta_t)}\right|^2 \tag{1.6}$$

$$T_p = |t_p|^2 = \left|\frac{2\,\sin\theta_t\cos\theta_i}{\sin(\theta_i + \theta_t)\cos(\theta_i - \theta_t)}\right|^2 \tag{1.7}$$

$$T_s = |t_s|^2 = \left| \frac{2 \sin \theta_t \cos \theta_i}{\sin (\theta_i + \theta_t)} \right|^2 \tag{1.8}$$

where θ_i is the angle of incidence of the light on the interface;
θ_t is the angle of the transmitted light as determined from Snell's law;
r_p is the P polarization reflectivity coefficient;
r_s is the S polarization reflectivity coefficient;
t_p is the P polarization transmission coefficient;
t_s is the S polarization transmission coefficient;
R_p is the amount of P-polarized light reflected;
R_s is the amount of S-polarized light reflected;
T_p is the amount of P-polarized light transmitted; and
T_s is the amount of S-polarized light transmitted.

The reflectivity curve in Figure 1.11 shows the importance of securing optical components such as lenses, filters, and dielectric mirrors (see 2.3.21, 2.3.13, and

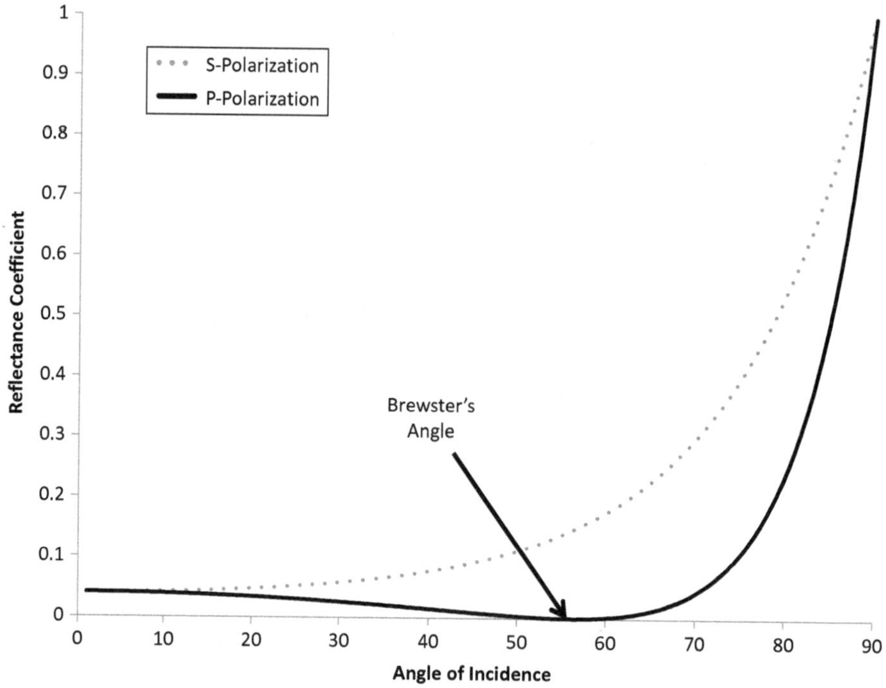

Figure 1.11

Reflectivity curve for interface between air (n = 1) and a piece of glass (n = 1.5).

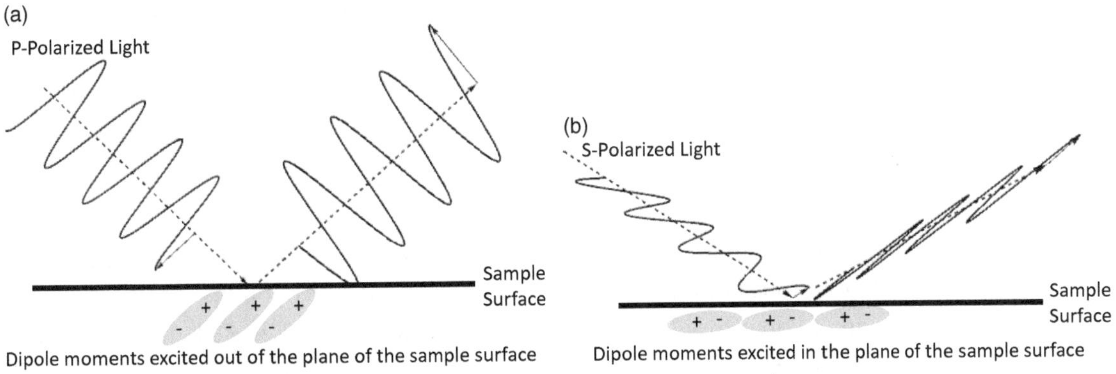

Figure 1.12

The dotted arrow represents the direction the light is propagating and the solid arrow represents the direction of the electric field. S- and P-polarized light reflect from a material differently because S-polarized light can only excite dipole moments in the plane of the surface the light is reflecting off. P-polarized light can excite dipole moments with a vector component that is perpendicular to the surface and a vector component that is in the plane of the surface.

2.3.18) at the proper angle; otherwise the polarization of light that reaches the sample may not be what you originally intended.

There is a special angle referred to as **Brewster's angle**, which is the angle at which only S-polarized light is reflected. Pieces of glass set at the Brewster angle are called **Brewster windows** and can be used to select for a specific polarization of light in an optical setup. Brewster windows are often used in the output coupler in gas lasers and semiconductor lasers, which is why those lasers are usually linearly polarized.

S- and P-polarized light can interact with a sample differently because S-polarized light can only excite dipole moments in the plane of the sample, while P-polarized light can excite dipole moments with vector components that are both in the plane of the sample and out of the plane of the sample (Figure 1.12).

1.8 Total Internal Reflection

There is a special case of reflection when light is traveling from a higher index material to a lower index material. Beyond a specific angle of incidence, all the light is reflected back into material 1 and no light propagates through the interface into material 2. The cutoff angle at which all of the light remains in material 1 is called the "**critical angle.**" This phenomenon is referred to as **total internal reflection (TIR)**. The critical angle only depends on the difference in the index of refraction between the two materials and can be solved for with Equation (1.9):

$$\theta_c = \sin^{-1}\frac{n_2}{n_1} \tag{1.9}$$

Total internal reflection forms the functional basis of fiber optic cables. Light propagates in a glass fiber with index of refraction of ~1.5 that is coated with a polymer with a lower index of refraction of ~1.3. As long as the light is injected into the fiber (i.e., "coupled") properly (i.e., at an angle below the critical angle), the light is totally internally reflected each time it hits the glass–polymer interface, and continues to propagate down the length of the fiber. We will discuss fiber optics further in Section 2.3.37, but this explains why defects in fiber optics can be detrimental to their efficiency in getting the same amount of light out that you put in; deviations in the angle of the internal glass wall with respect to the propagation direction of the light (i.e., the long-axis of the fiber) at defects in the fiber can cause some light to be lost to the outer polymer sheath since the condition for total internal reflection no longer holds. This is one of several reasons to be gentle while handling fiber optic cables.

An important detail about TIR is that even though no energy propagates into material 2, some of the electric field of light does penetrate into material 2 and can interact with molecules in material 2. The electric field at the interface switches

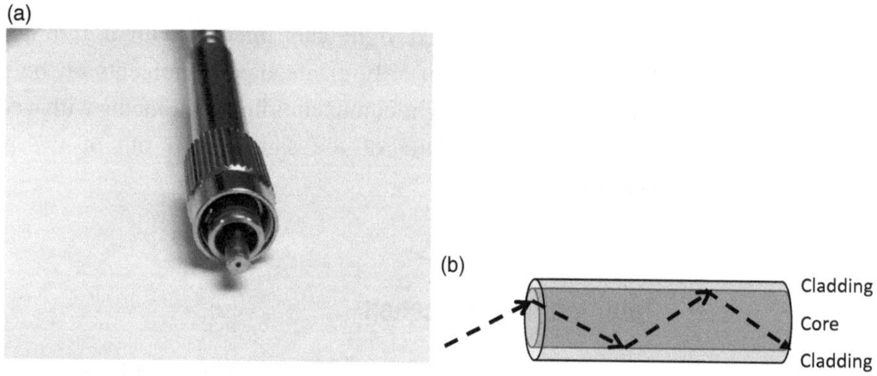

Figure 1.13

(a) End of a fiber optic cable. The glass fiber is in the very center of the tip. There is a polymer cladding around the glass fiber and then a metal jacket to help protect the fiber from being accidentally broken. (b) The cladding must have an index of refraction lower than the core for total internal reflection to occur at the interface.

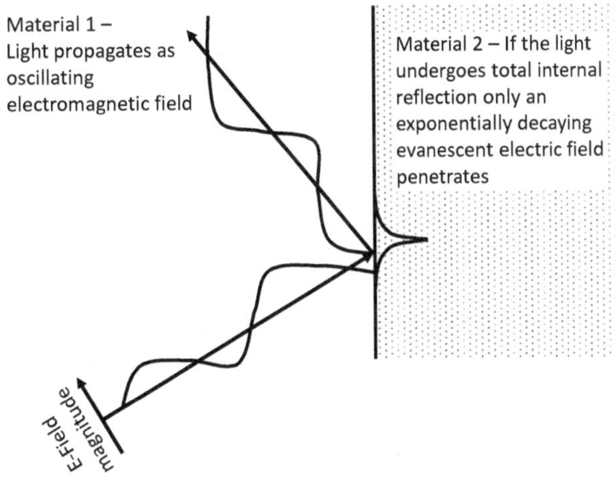

Figure 1.14

When light undergoes total internal reflection, all the light is reflected back from the interface and only an exponentially decaying evanescent electric field penetrates into material 2. The penetration depth depends on the wavelength of light and the indices of refraction of the two materials but it is typically several hundred nanometers.

from an oscillating sinusoidal electric field to an exponentially decaying electric field. The field that penetrates beyond the material 1/material 2 interface is known as the **evanescent electric field**. This phenomenon forms the basis of several imaging and spectroscopy techniques (e.g., TIRF, ATR-FTIR). The details of how these techniques take advantage of the evanescent electric field will be discussed in detail in Sections 3.4.2 and 4.8.

The reason that the evanescent electric field exists is that nature does not allow for discontinuities in electric fields. If the electric field decayed instantaneously to zero at the boundary, then the electric field would simultaneously be finite and zero at the interface – a physical impossibility.

1.9 Absorption

Photons have no mass so they can be created and destroyed at will provided that their energy is conserved. The energy that a photon contains can be used to move an electron in a material from a ground energy state to an excited energy state. The photon is destroyed in this process. This is known as **absorption**. (An IR photon can also move a molecule from a vibrational ground state to a vibrational excited state, and extremely high-energy (>MeV) photons can interact in other ways with the nuclei of atoms, though this high-energy regime is outside the scope of this book.)

By carefully measuring the energy of the photons that are absorbed and the energy of the photons that are emitted, a material's energy levels can be deduced. This is the basis for UV-Vis spectroscopy, FTIR spectroscopy, and fluorescence (photoluminescence) spectroscopy, all of which are discussed in Chapter 3.

1.10 Scattering

Light can also be scattered when it interacts with a material. In scattering events, the propagation direction of the light is not preserved. Photons can be scattered **elastically**, where they don't lose any energy during the scattering event, or **inelastically**, where they do lose some energy – typically due to coupling to vibrational motion of a molecule in the material. Elastic scattering is known as Rayleigh scattering. Inelastic scattering is known as Raman scattering (Section 3.5). It is important to note that light scattering is a different physical process from absorption and emission phenomena.

1.11 Emission

When the electron returns to the ground energy state, a photon is created with the exact same energy as the energy difference between the excited and ground energy levels. This is known as **emission**, **fluorescence**, or **luminescence**.

Electrons can be excited to a higher energy state from which they can emit photons as they relax back down by a number of means. The light emitted by the different excitation methods will contain different information about the sample. Some common means of excitation are:

photoluminescence – photon excitation;
triboluminescence – friction excitation;
thermal radiation – heat excitation;
electroluminescence – electrical excitation;
cathodoluminescence – electron beam excitation in an SEM or TEM; and
chemiluminescence – chemical reaction excitation.

An incandescent light bulb emits light by thermal radiation (Appendix 4). Fluorescent light bulbs and some white light LEDs create their white light by a combination of electroluminescence and photoluminescence phenomena. It is important to note that even though all three light sources look white to the human eye, they produce very different spectra. The human eye has four types of photo receptors. There are rods and three types of cones. The rods are responsible for low-light vision and have no color discrimination. The three types of cones are each responsive to red light, green light, and blue light. As long as a red-sensitive cone, a green-sensitive cone, and a blue-sensitive cone are stimulated, the human brain interprets the color as white.

The incandescent light bulb produces a full continuum of wavelengths, known as blackbody radiation (Appendix 4), that is weighted more heavily to the red. The ratio of red photons to blue photons can tell you how hot an object is. A "red hot" object is cooler than a "white hot" object. As an object heats up there are more blue photons, as there are more electrons in higher-energy states. The continuum of energies on photons emitted tells you that there is a continuum of electronic energy levels between which the electrons can transition.

A fluorescent light bulb works by electrically exciting a plasma of gas inside a tube. The emission lines from the gas are sharp and discrete. The photons come out at specific wavelengths. The specific wavelengths tell you what the atomic compositions of the plasma in the tube are. There are discrete atomic energy levels between which the electrons can transition. Manufacturers will try to pick a mix of gases that will stimulate the red-sensitive, green-sensitive, and blue-sensitive cones in the eye in a ratio that is perceived as pleasant. Sometimes the tube of the bulb is coated with a phosphor coating that will absorb some of the blue or ultraviolet light and re-emit in the red to give a "warmer" color to the light.

LED stands for light emitting diode (Section 2.2.4). An LED is a semiconductor device that allows electrons provided by an electrical circuit to drop across a specific energy level at a specific junction in the semiconductor material and produce a photon. A white light LED can be three independent LEDs that produce red light, green light, and blue light, or a single blue LED with phosphor coatings that will absorb some of the blue light and re-emit in the green and red spectral regions. The emission spectrum from an LED contains information about the energy levels in the semiconductor material and phosphors which would provide clues to their chemical identity.

1.12 Birefringence

As light propagates through a material, the electric field of the light interacts with the electrons that compose the chemical bonds in the material. The chemical bonds in a material may not be rotationally symmetric. Birefringence occurs when light propagates through a material with two different indices of refraction along two different sample orientations. In birefringent materials, incident light can be split into two optical paths because of the two indices of refraction.

A classic example of birefringence is the double image observed through a piece of optical calcite (Figure 1.16). The crystal has two different indices of refraction along the two different orientations of the bonds, also known as crystal axes. If we apply Snell's law but with two different indices of refraction for the two different crystal orientations, we can understand why there are two images. In fact, the two different images are of orthogonal polarization to each other. Calcite crystals are used to make high-quality polarizers for lasers exploiting this phenomenon. The beam that passes straight through the crystal is referred to as the **"ordinary" beam** (often denoted "o"), while the refracted beam is referred to as the **"extraordinary" beam** (often denoted "e").

Many other materials also exhibit birefringence. For example, stress in a material will cause **anisotropic** (i.e., spatially non-uniform) bonding arrangements. For this reason, plastics often exhibit birefringence because the pulling process used to manufacture many hard plastics yields anisotropic bonds. Observing the birefringence of a material between two crossed polarizers allows one to image the mechanical stress in a transparent sample (Figure 1.17). The birefringence of materials can be a problem in some microscopy techniques such as differential interference contrast (DIC; Section 4.6).

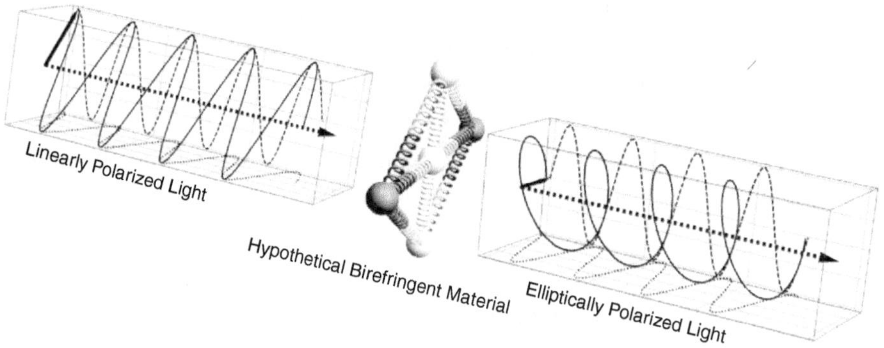

Figure 1.15

The bonds in the birefringent material are not isotropic. It will be easier for the *y* component of the light ray to propagate as compared to the *x* component in this hypothetical example. This will lead to a phase delay between the *x* and *y* vector components. This phase delay between the vector components will change the polarization of the light from linear to elliptical. Depending on the orientation of the crystal, two spatially separated beams with orthogonal polarization can emerge from the other side of the crystal.

Figure 1.16

Double image through optical calcite crystal demonstrating the birefringent properties of the crystal. The two images are made up of orthogonally polarized light. The chemical formula for calcite is $CaCO_3$.

Figure 1.17

The mechanical stress in the plastic petri dish induces different amounts of birefringence. This can be imaged by placing the plastic dish between two crossed polarizers.

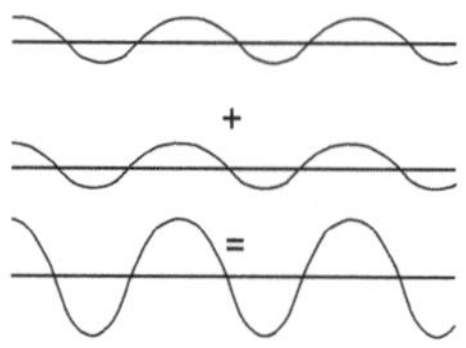

Figure 1.18

When two waves are in phase, the peaks and the troughs occur at the same time; they can add constructively, leading to a bigger wave.

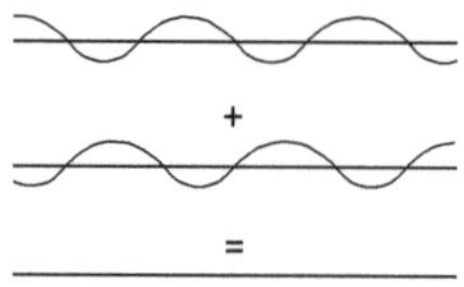

Figure 1.19

When two waves are out of phase, the peaks of one occur at the troughs of the other; the waves destructively interfere leading to no resulting wave.

1.13 Interference

Two light waves can interact by interference. Interference is the way that the amplitude of two waves add together. In the case of two waves whose peaks and troughs are aligned, they undergo **constructive interference**, in which the amplitudes of the waves add together to yield a wave with higher amplitude (Figure 1.18). **Destructive interference** occurs when two waves have peaks and troughs that are not aligned; the resulting wave has lower amplitude (Figure 1.19). Waves with different wavelengths can both constructively and destructively interfere at different positions.

Typically, light waves need to have similar frequencies and spatial locations to interfere. For instance, radio waves (with wavelengths of several meters) usually don't interfere with visible light waves (with wavelengths of 400–800 nm).

1.14 Ray Optics

One of the simplest and most common techniques to understand light uses arrows to describe the path that light takes through a medium such as air, water, your optical setup, or some combination of materials. This technique is known as **ray tracing**; **ray optics** is the term that optics experts use to describe analysis of a system by considering the paths in which photons move. Consider the following examples.

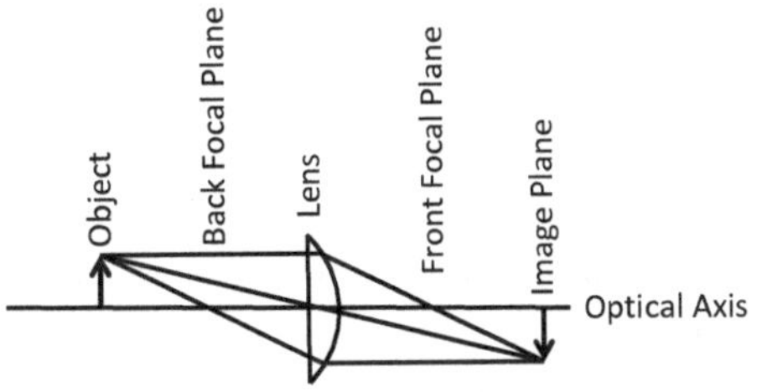

Figure 1.20

Ray tracing for a single lens.

A photon moving through a single medium will not change direction. Therefore, it is depicted as a straight arrow.

As discussed previously, a photon that approaches an interface will bend into the medium with higher refractive index according to Snell's law.

Finally, two photons moving parallel to one another (known as a **collimated** beam) incident on a lens will be focused to a single **focal point** by the lens (more on lenses in Section 2.3.21). This is depicted as two parallel lines hitting a lens, and then bending toward an intersection point some distance from the lens.

A key concept that emerges from ray optics is the **optical axis**. The optical axis is the axis along which light travels through an apparatus or setup.

Ray tracing for a lens follows three simple rules:

1 A ray passing through the center of a lens will pass straight through.
2 A ray traveling parallel to the optical axis will cross the optical axis at the front focal plane.
3 A ray crossing the optical axis at the back focal plane will emerge parallel to the optical axis.

1.15 Real Versus Reciprocal Space

As we mentioned earlier, light carries information. The information that the light carries is arranged differently at different points in the optical path when you are going through focus. If you look at Figure 1.20, you will notice that the image plane and the focal plane are actually in different positions. The light carries the same information about the object at both positions, but they can be thought of as plotted in different units at the two different positions. This is a slightly counterintuitive concept and has a number of different terms applied to it. The information at the image plane can be thought of as being plotted in **real space**, as the information in x and y of the image have a clear one-to-one relation with the real sample and are plotted in units of length. The information at the focal plane is said to be arranged in **reciprocal space**, **Fourier space**, or **k-space**. The information can be thought to be the Fourier transform of the real spatial dimensions and to be plotted in spatial frequencies. Spatial frequencies have units of 1/length. (Transmission electron microscopists routinely switch back and forth between the image plane and the focal plane to take advantage of this property. In the image plane, they have an image of their sample. In the focal plane, they have a diffraction pattern of their sample that clearly shows information about the crystallinity of their sample.)

Spatial filters can be positioned at the focal plane to manipulate what spatial frequencies are transmitted. This field of optics is known as **Fourier optics**. Modern microscope objectives are designed so that the back focal plane is the last surface of the microscope objective. This means that the microscope objective itself acts as a low-pass filter for which spatial frequencies will contribute to the image formation. Fourier optics will be further discussed in its application to structured illumination microscopy in Section 4.16.

1.16 Further Reading on Optics

Optics by Eugene Hecht is a classic undergraduate text on optics presented from a physics standpoint. It is an excellent reference and contains many of the mathematical underpinnings for the concepts that were discussed here qualitatively.

Principles of Optics by Born and Wolf is more a graduate-level text that delves deeper into the mathematical descriptions of optical phenomena.

2 Introduction to Common Optical Components

In this chapter, we will discuss the most commonly encountered components in working with optics. These are the tools with which you manipulate light to extract the information about the sample that you are interested in.

2.1 Light Safety

First, we want to cover the most important aspect of optics lab work:

Safety

The single biggest thing you can do to stay safe is to wear appropriate laser safety goggles. The laser jock's black humor mantra is "Don't look into the laser with your one remaining good eye."

In order to choose appropriate laser goggles the first question to answer is: "What wavelength of light will you be working with?" Not all goggles block the same wavelengths, or provide the same protection. Most laser goggles have a label along the top edge that specifies what wavelengths are blocked, along with the degree of blocking power in units of optical density (OD). Optical density is a log scale. OD of 1 means that 10 percent of the incident light is transmitted; OD of 2 means only 1 percent of the incident light is transmitted. For most laser applications, OD 6–7 is appropriate. Each system is a little different; consult with your institution's laser safety officer.

Situational awareness is also extremely important in a laser lab. It is important to understand where the beam path is before you turn on the laser. Remember that you won't be able to see the laser beam when it is on, because you will be wearing laser safety goggles that prevent the laser light from reaching your eyes. One of the biggest risks when working with lasers is accidentally putting something reflective into the beam path and causing a reflection that sends the beam in an unexpected direction. Knowing where the beam is at all times is key to preventing accidents. Additionally, placing "beam blocks" in or to the side of the laser path is an important part of controlling where the laser is going. Beam blocks can be made

Figure 2.1

Laser safety goggles. The wavelength and degree of blocking are usually written on the top edge of the laser goggles.

Figure 2.2

A piece of paper or cardboard taped to a post holder can serve as a simple beam block for low-power lasers.

easily from pieces of cardboard taped to optical bases or posts. Be careful though: high-power lasers will ignite cardboard and paper, so sometimes specialized metallic beam blocks or beam dumps are needed.

TIP: Drawing the beam path directly on the optics table with a Sharpie marker is a useful way to remind yourself and others where the laser beam path is. A little acetone on a paper towel will take the Sharpie line off the stainless-steel table when you need to change the setup.

The eternal dilemma when working with lasers is that the safety goggles that are carefully designed to prevent the laser from entering your eye also prevent you from seeing where the laser beam is going. There are a variety of ways to image a laser beam while wearing goggles, so please fight the urge to peek over the goggles. Another important point is that laser beams are sometimes in the UV or IR spectral region and you can't see them with your bare eyes anyway.

Most optics suppliers sell viewing cards that when placed into the laser beam will shift the wavelength of light to a range that can pass through the laser goggles. These cards can work by either fluorescence or second harmonic generation. For IR lasers a liquid crystal sheet or an IR viewer camera can be used to trace the beam. Infrared cameras are now even available as cell phone attachments (see Section 2.4.3).

For pulsed lasers, "burn paper" is sometimes useful. Burn paper is basically low-sensitivity photographic paper that will darken with exposure to the high intensity laser beam.

When the beam path for a setup is done, including a "guide beam," (a low-power beam that is of a wavelength visible through laser goggles, and that is collinear with the invisible beam) is useful.

TIP: The neon-colored index cards from an office supply store can often be used as a cheap laser viewer card. Try different colors to see which one works best for your laser and safety goggle combination.

When entering an optics lab, you should remove all reflective objects (watches, rings, necklaces, etc.) that might accidentally end up in a beam, causing the laser to be inadvertently redirected. Check around the lab for things that might cause unintentional reflections if the beam gets away from you, such as glass doors, cabinet windows, monitor screens, cups and glasses, metal tweezers, screw-drivers, etc.

TIP: Covering up reflective surfaces in the lab (e.g., glass cabinet doors) with some black construction paper using black electrical tape is a wise precaution.

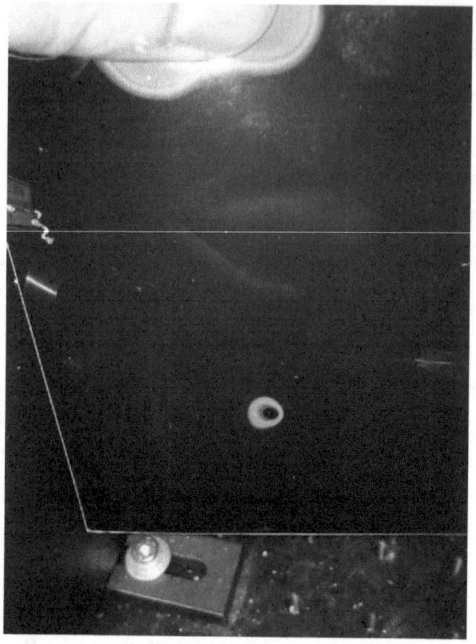

Figure 2.3

Liquid crystal sheets are sometimes a good solution for viewing infrared laser beams. The IR laser will locally heat a spot on the liquid crystal. The liquid crystal sheet can be cut into convenient shapes to put into the beam. Do be careful that the beam is not too powerful and simply burns the liquid crystal sheet.

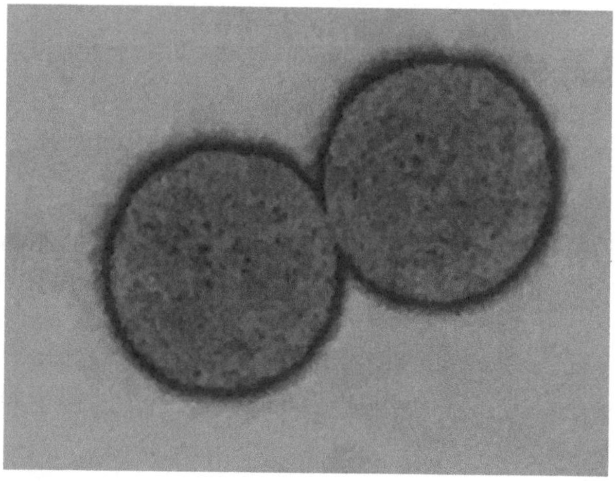

Figure 2.4

Burn paper allows you to characterize the quality of the pulsed laser beam. Two burn spots show nice Gaussian shaped beams. A bad beam profile would have hotspots not at the center of the beam.

Setting up a new beam path is the most dangerous part of working with an optical setup, since this is when the beam is most likely to be deflected in an unexpected direction as you move mirrors and other optics around. When setting up a new experiment, always lay out the beam path on paper first, and try to assemble and position the optics such that laser beams stay in the plane of the optics table. Sometimes setting up a low power co-linear "guide beam" first and checking the optical element alignment is a safer way to proceed. While aligning and building new setups, always try to work with the minimum laser power necessary to see what you are doing when setting up a new optical path.

TIP: Try to minimize the number of people in the room when you are setting up a new beam line.

In addition to eye damage, laser beams can burn skin and tissue (hence laser hair removal clinics). Generally, the burn threshold for skin is a couple hundred milliwatts, but this depends on the wavelength range used and the pulse rate of the laser.

TIP: Wearing long sleeves and disposable nitrile gloves in the lab is a good idea. Be careful about droopy sleeves though, as these might inadvertently slip into the beam line as you try to reach over it.

Some lasers are more dangerous than others. All lasers are categorized by a class which denotes their danger level. Class 1 is the least dangerous and class 4 is the most dangerous. (Sometimes the classes are written in Roman numerals I, II, III, IV.) Most lasers have a sticker on them near the exit port stating what class they are.

The potential hazards of lasers are classified by the following labels:

Class 1 lasers typically have a fully enclosed beam or very low power levels and are safe under normal operating conditions and will not cause harm to your eye.

Class 1M lasers are safe except when viewed through some concentrating or magnifying lens such as a telescope or camera viewfinder.

Class 2 lasers are typically safe because their power is low enough (<1 mW) that they will not harm your eye before you blink and look away, which for most people takes about a quarter of a second. However, extended exposure is almost guaranteed to cause harm to your eye. Most laser pointers are class 2 lasers.

Class 3R lasers have powers of 1–5 mW in the visible range, and could potentially cause harm to your eye in the time it takes you to blink or look away. Some very bright laser pointers are classified as 3R.

Class 3b lasers, which for visible light include any laser with power between ~5 and 500 mW, are harmful if the direct laser beam enters the eye. However, diffuse or scattered light is generally not harmful (but it's good practice to prevent it from getting into your eyes anyway by simply always wearing your laser goggles). Because 3b lasers cause harm, they must be interlocked (with, for instance, a key) so that they cannot be turned on without considerable forethought and effort.

Class 4 lasers are dangerous – they will burn your skin and cause permanent damage to your eyes. Their high power also makes them a fire hazard, so be sure to keep anything flammable away from the beam paths of these lasers! Reflections and scattered light from these lasers are usually dangerous, so goggles should *always* be worn any time these types of lasers are enabled.

In addition to goggles, physical or lab-based safety features, known as "Engineering Controls," can help to prevent laser-related injuries. Examples include the following:

- Plastic tubes around high power laser beams so you can't reach into them accidentally (especially for class 4 lasers!).
- "LASER IN USE" signs so that your unwitting coworkers don't walk into the laser lab while the laser is on. Be sure to actually turn off the "LASER IN USE" sign when you are done so that people actually pay attention to it and take it seriously. If it is on all the time people will start to ignore it.
- Electrical interlocks so that you can't turn on the laser without an extra, deliberate button push.
- Black curtains and walls to contain laser light.
- Black beam blocks placed to the side of the laser path, so that stray in-plane reflections do not reach you or reflective or flammable objects throughout the rest of the lab.

Also please be aware that the laser enclosures and other support accessories often pose additional safety hazards. First, some laser power supplies operate at very high voltages (>kV) or very high currents! Do not try to repair a power supply yourself. Be careful that you enclose any power supplies or wires in sufficiently insulating boxes such that a wayward grab in the dark for a screwdriver won't accidentally result in an electrical shock! If you have questions about a laser power supply, consult the manufacturer or a very qualified and experienced laser electronics expert. Many gas lasers present specific safety risks. Since these

Figure 2.5

Laser curtains and an illuminated "LASER IN USE" sign

lasers contain tubes that are often partially evacuated or at high gas pressure, small cracks from rough transport or striking the tubes with rigid tools can yield tube explosions or implosions. For this reason, most gas laser tubes are enclosed in a mesh and an additional metal box to prevent pieces from flying around the lab. The majority of gas lasers are filled with inert noble gases such as helium, neon, argon, and krypton. But, some more sophisticated systems contain halogens such as fluorine and chlorine gas, which are toxic. Finally, with a water-cooled laser, think about where the water will pool if there is a leak. Please make sure that leaking water won't enter the high-voltage power supply.

Laser safety tips and tricks review:

- ***Always*** wear your safety goggles.
- Don't ever get comfortable with your setup. Experienced laser users have more accidents than beginners because they get too relaxed.
- Be careful with your safety goggles, too! Scratches on the goggles can scatter light and allow it through the otherwise-impenetrable coatings (see cleaning optics in Section 5.1).
- Never lean over or crouch down to pick things up off the floor, never bend over to tie your shoelaces, or in any way put your eye or face at beam/table level. This is the easiest way to prevent injury.

- Take off your watch and other reflective metal jewelry before working with dangerous light sources.
- Keep track of stray and upward reflections and extra beam paths. Use apertures and beam blocks to deal with these (see Sections 2.3.19 and 2.3.20).
- Be careful when touching any optics (especially beam blocks) that have been exposed to >0.2 W of power as they might be hot.
- Don't let shiny (screwdrivers, extra optics mounts, tweezers, clutter, samples) or flammable (solvent bottles, lab notebooks, paper, samples) objects accumulate near your setup.
- If you have long hair, tie it back so that it does not fall into the beam path as you move around. (Note again: don't bend over and bring your eye level with the beam path.)
- Put "engineering controls" in place.
- Be aware that lasers pose hazards in addition to eye damage, including a burn and fire hazard and an electrical hazard.
- Talk to your institution's laser safety officer. They're around to help you. Often times, they will even purchase goggles and materials for engineering controls for you, which can save you a lot of time and your lab a lot of money.
- Ask the company that sold you the laser for help on safety and which goggles they recommend. They had to test it in their lab at one point, so they'll have good suggestions on how to operate the laser safely.

2.2 Light Sources

When we say light source, we mean anything that emits light in your lab. Examples range from **lasers** to **light bulbs** to **LEDs**. Light sources such as the sun, computer screens, and your cell phone might also count as light sources in the lab and can influence your measurement if you're not careful.

TIP: Taking the time to run the control experiments of characterizing the room lights emission spectrum, the computer screen emission spectrum, your cell phone screen emission spectrum, etc. can save you time in chasing down erroneous signals later. If you have a choice, overhead LED light fixtures have fewer sharp spectral features and are thus a better choice for spectroscopy labs. Overhead fluorescent lights have a number of sharp peaks that can be misleading. The line at 612 nm from fluorescent lights is a particularly strong and clean peak.

2.2.1 White Light Sources

As mentioned previously, human eyes have three color receptors (red, green, and blue). When all three receptors are stimulated, your brain perceives this as white light. It is important to note that not all white light sources actually emit all the wavelengths of visible light. A computer screen, for instance, can display white light by turning on red, green, and blue pixels with even intensity. The screen does not emit yellow, orange, or violet light. "White" LEDs use the same trick either with a package of three LEDs (red, green, and blue) combined or one blue LED with a phosphor coating that absorbs a portion of the blue light and re-emits in the green and red. It is important to keep in mind that not all white light sources have the same spectral profile, so you may need to plan carefully when designing your experiment to achieve your desired spectral profile.

If you do need a white light source that emits a continuous spectral profile, metal halide lamps (also known as halogen lamps) are a cheap and easy option. These bulbs consist of a tungsten filament that is heated by running electrical current through it until it glows white hot. It has a characteristic **blackbody radiation** curve, but it does contain all the wavelengths of light. They generally operate at high temperatures and pose burn and fire hazards, so make sure some engineering controls are in place to prevent accidentally touching them with your hands and stray paper from landing on the lamps. Also, don't touch these bulbs with your bare hands since the oils from your skin can weaken the glass or quartz, leading to a break.

Xenon lamps contain a tube of xenon plasma that does a reasonable job of providing all the wavelengths of light, but will have some much stronger plasma lines. Mercury lamps are another white light option that is falling out of fashion due to environmental concerns, but are still widely used. They emit mostly at specific plasma lines spread throughout the UV and visible range, giving the perception of white light. Xenon and mercury lamps do emit a significant portion of UV and near IR light that should be considered. Therefore, when working with xenon or mercury lamps, UV-blocking safety goggles are essential. Also caution should be used with xenon or mercury lamps as they tend to become quite hot when operated. Note that xenon and mercury lamp housings for microscopes often have UV and/or near IR filters to cut out the wavelengths that the manufacturer assumes you will not want hitting your samples. If you are doing fluorescent microscopy of live cells, this is probably true as you do not want to cook your sample while you are looking at it. If you are trying to use the microscope for materials science applications, it is worth keeping in mind that you may not get all the wavelengths that you might be expecting due to the built-in filters.

Supercontinuum lasers offer significantly higher power white light than can be achieved with a bulb or diode, but their cost is correspondingly higher as well. In contrast to the pumped lasers in 2.2.3 that use a single physical process to convert a pump laser beam into a narrow wavelength emitted beam, supercontinuum lasers rely on a number of nonlinear processes all occurring simultaneously in a single cavity to convert an input pulse into output spanning a fairly wide wavelength range. For this reason, supercontinuum lasers are inherently pulsed. (Pulsed lasers as compared to continuous wave lasers are discussed in Section 2.2.5.) Speak with manufacturers about specific needs, as outputs can be customized per application. Given the wide range of emission from supercontinuum lasers, be sure to carefully consider the wavelength range for your laser safety goggles.

2.2.2 Synchrotrons

Synchrotrons are particle accelerators that shoot electron beams through a curved kilometer-scale vacuum tube. As the electrons change direction as they move around the curves, they lose energy, which is radiated as high-energy photons. The number of photons is proportional to the number of electrons traveling through the synchrotron. The electron beams can be extremely powerful. Therefore, synchrotrons offer high-energy (x-ray) as well as extremely high-brightness (i.e., lots of photons) light sources of different wavelengths from x-ray to IR.

The high-brightness beams offer higher signal-to-noise ratios and higher resolutions than those afforded by bench-top setups. Synchrotrons also allow users to test samples that have low volumes and to collect data faster than one could in a typical lab setup. Typical applications of synchrotron radiation are x-ray crystallography, ultrafast spectroscopy for characterizing ultrafast molecular and protein dynamics, high-resolution and other specialized lithography, high signal-to-noise x-ray fluorescence and photoelectron spectroscopy, high-resolution x-ray tomography, and IR tomography. Because of their size and complexity, synchrotrons are maintained by universities and national labs (e.g., Brookhaven National Lab, Argonne National Lab, the Advanced Photon Source, etc.) that offer outside users "beam time." While many techniques discussed in this book can be transferred to synchrotron beam lines to take advantage of the high brightness of the light source, details of synchrotron experiments are beyond the scope of this book.

2.2.3 Lasers

What is a laser? LASER is an acronym that stands for Light Amplification by Stimulated Emission of Radiation.

Laser light has several convenient properties. It is coherent, meaning that all the photons in the beam are in sync and doing the same thing. Laser light can be formed into a beam that can propagate without diverging. This means that the laser beam diameter is the same near the source as on the other side of the lab. A normal flashlight by comparison has a divergent beam, meaning that the beam gets bigger the farther you are away from the source.

Laser light is often monochromatic. In the wave picture of light that means the photons all have exactly the same wavelength. In the particle picture of light, every photon in the beam has exactly the same amount of energy. Laser light follows the normal rules of light discussed in Chapter 1: it can interfere with other light, can be focused by a lens, has a polarization, etc.

When an electron in an ion or molecule transitions from a higher energy level to a lower energy level, it can create a photon to carry away the excess energy. This process of photon creation is fluorescence or luminescence. If we can set up a system where all the photons are doing exactly the same thing at the same time, we have a "laser."

In its simplest form, a laser consists of a lasing medium (aka "gain" medium) and two mirrors on either side of the lasing medium. This geometry is known as a **resonant cavity** or **laser cavity**. One of the mirrors at the ends of the cavity is slightly transmissive, so a small fraction of the light can leak out of one end of the cavity; this light forms your laser beam. The lasing medium is maintained in an excited electronic state by **pumping** with either other light or electricity. As electrons in the lasing medium relax, they emit photons, which can reflect off of the two mirrors at the ends of the cavity, sending the light emitted from the lasing medium back and forth through the medium. These photons can stimulate further emission of photons by the lasing medium on one of their many trips back and forth. Repeated again and again, this process yields a positive feedback loop. This positive feedback loop eventually synchronizes many of the properties of the photons, transforming the light from "fluorescence" (light going in all directions) to "lasing" (a coherent beam of light going in a single direction). Lasers are usually described by their lasing medium, as the lasing medium determines most of the important properties of a particular laser system. Lasing mediums can be gas, liquid, or solid, and can comprise ions, elements, compounds, or molecules.

Continuous wave lasers (commonly referred to as **CW lasers**) deliver continuous beams of light. The power (i.e., energy per second) of CW lasers is measured in **watts**, abbreviated W ($W = J/s = N\ m/s = kg\ m^2/s^3$). Typical powers you might encounter in the lab range from μW to hundreds of mW. Powers of >1 W are used often as energy inputs for other lasers (see e.g., DPSS lasers or pulsed lasers [Section 2.2.5]).

In contrast to CW lasers, **pulsed lasers** emit light in discrete time packets. Note that pulsed lasers are fundamentally different from **chopped** CW lasers. If you

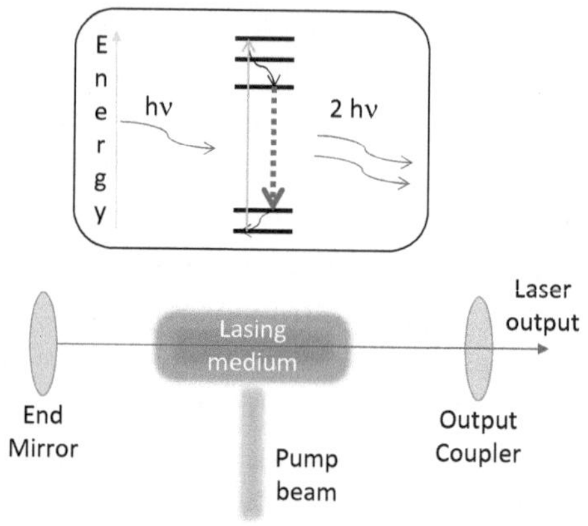

Figure 2.6

A laser in its simplest form is comprised of two mirrors and a lasing medium in a cavity. The lasing medium is a fluorescent material that is constantly pumped into an excited electronic state. The two mirrors provide a positive feedback mechanism which synchronizes all the properties of the photons in the laser beam.

chop a CW beam (basically by putting a rotating beam block in the beam path), you are throwing away energy. In a pulsed beam the total energy is redistributed to arrive at some times and not at others. Pulsed laser beams have very high peak powers even if the average power is the same as a CW beam.

Pulsed light is indispensable for nonlinear optics, various forms of optical pumping such as pulsed laser deposition, and studying the dynamics of processes which occur on extremely fast time scales such as molecular bond vibrations, electronic transitions, and chemical reactions. Pulse lengths that you will encounter in most labs are nanoseconds (ns; 10^{-9} s), picoseconds (ps; 10^{-12} s), or femtoseconds (fs; 10^{-15} s). To propagate, a light wave needs to complete one full cycle of the electromagnetic wave. In the visible light spectral range this corresponds to about 5 fs being the shortest possible pulse. Attosecond (as; 10^{-18} s) pulses are obtainable now, but fairly uncommon at the moment. The wavelength of an attosecond pulse is typically in the soft x-ray region of the electromagnetic spectrum which presents its own range of hazards and safety concerns. Commercial table top systems are now available from KMLabs.

Gas and metal vapor lasers are common types of CW lasers found in research labs. Examples include helium–neon (HeNe – often pronounced "he-nee"), argon (Ar^+), krypton (Kr^+), and helium–cadmium ($HeCd^+$). Gas lasers operate by applying an extremely high voltage across a glass tube full of a gas mixture (e.g., He and Ne) to excite specific electronic transitions in the gas. As electrons relax, they emit at wavelengths characteristic of the electronic transition in the gas molecule; each type of gas has unique emission wavelengths, which are known as

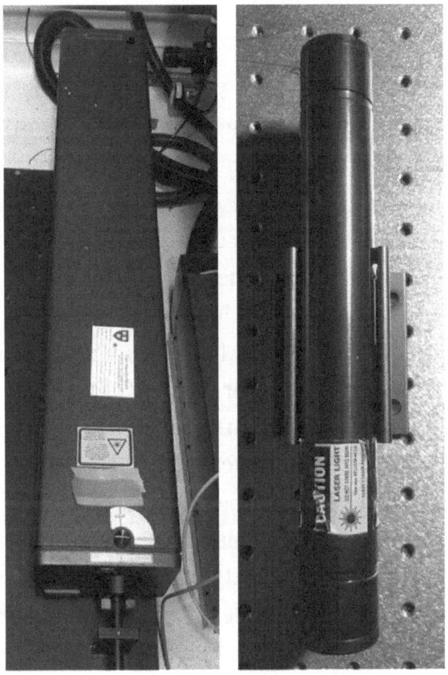

Figure 2.7

Gas lasers can be large or small. On the left a mixed gas krypton–argon laser. On the right a helium–neon laser.

plasma lines. Most gases emit at multiple plasma lines. The unwanted emission wavelengths can be removed with optical filters. The optical path of the laser system is set up so that one (or more) of these lines effectively ends up in a positive feedback loop, leading to "lasing." The seemingly arbitrary values of the "standard" laser lines (i.e., 488 nm, 514 nm, 633 nm, etc.) are inherited from the material used as a lasing medium. Advantages of gas lasers include low beam divergence, high optical power densities, and very narrow spectral **linewidth**, meaning that the light emitted from the laser spans less than 1–2 nanometers in wavelength. However, because these lasers require high-voltage power supplies and special gas tubes, they can be quite expensive and require significant cooling via fans or circulating water. It is important to isolate these vibrating cooling systems away from vibration-sensitive measurements.

TIP: Putting power supplies and water chillers either on the floor below the optics table or on a mechanically isolated shelf above the optics table helps isolate vibrations.

Liquid dye lasers were the first tunable laser and are starting to come back into use after falling out of favor. One of the major advantages to dye lasers is that their wavelength can be tuned over a fairly wide range. A major disadvantage of dye lasers is that the dyes used as a gain medium are typically quite carcinogenic. Solid state optical systems such as optical parametric oscillators (OPOs), or "white light lasers," can also offer wavelength tunability.

Solid state lasers come in several forms. In some, semiconductors comprise the gain medium and the laser functionality stems from the recombination of electrons and holes at a p–n junction. Other solid state lasers have gain media comprising optical crystals that hold ions with desirable electronic transitions in place such as titanium sapphire or Nd:YAG lasers.

Diode and **quantum well** lasers, which operate by recombining electrons and holes in semiconductor crystals, are becoming increasingly common for a variety of wavelengths. These lasers are very small, only require a standard lab power supply to apply small voltages (<10 V) and currents (~100 s of mA) until you need very high (>800 mW) output powers, and are often quite affordable, though blue and UV diode lasers are much more expensive than green, red, and IR equivalents. Cost is determined by the diode's output wavelength, output power, and the **divergence angle** of the light coming out of the diode and can range from $50 to tens of thousands of dollars. Poor beam divergence and linewidth, especially at higher output powers, are major disadvantages of laser diodes; the cone angle of the beam exiting the diode can be as high as 30 degrees, meaning that careful compensation with lenses is required to "**collimate**" the beam, or transform it into a parallel beam. Laser diode wavelength is also somewhat temperature dependent, so diode lasers are often quoted with a nominal wavelength ±5 nm. This may or may not make a difference to your experiment, but it is something to keep in mind. **Quantum cascade lasers (QCL)**, a type of quantum well laser which has multiple quantum wells in a single semiconductor crystal, are useful for mid-IR emission. They are highly tunable over a narrow spectral range.

Diode-pumped solid state (DPSS) lasers are a newer type of CW light source which contain two light sources. First, a diode laser emits light with wavelength usually at 800–900 nm. The light from the diode laser is then used to pump a solid state crystal to the point of lasing. The solid state crystal and its impurities determine the wavelength of the laser; the most common crystal is Nd:YAG (commonly pronounced as "enn-dee-yag"). Much like the plasma lines of gas lasers, Nd ions have unique electronic transitions; the strongest transition emits at 1064 nm and weaker emission lines are 914 and 1342 nm. These outputs can then be frequency multiplied or frequency added to various green, yellow, blue, and UV wavelengths. Compared to diode lasers, DPSS lasers are spectrally much more stable. However, they are often larger and more expensive. Optically pumped semiconductor lasers (OPSLs) are a subset of DPSS lasers – in this case, the first laser pumps a second diode.

2.2.4 Light Emitting Diodes

If you do not need high power or a highly collimated beam, **LEDs** are a good alternative for illumination – they operate generally by the same principle as laser diodes, come in a variety of colors (including IR and UV), and only cost a few dollars. Just be careful not to apply too much current to them or run them in reverse polarity, as most LEDs will burn out.

For both diode lasers and LEDs, manufacturer and supplier specification sheets provide plots showing the intensity of the light emitted vs. the amount of current injected into the diode.

TIP: You can often see a black spot inside the LED casing, hear a small fizzle or pop, and smell a bit of burning plastic if an LED is burned out by running it in reverse polarity or too much current is pushed through it. Consult the specification sheets for LEDs to determine how much current an LED can tolerate before burning out.

TIP: Small LEDs often have two leads, or legs. The longer leg corresponds to the one that should face the positive side of the power supply. Some LEDs, such as multicolor diodes, have three or four legs; you can consult the specification sheet for their connection functionality and wiring diagrams.

2.2.5 Pulsed Lasers

To create a pulsed laser, one could naively open and close a shutter in front of a CW laser. However, that would be a **chopped** laser and not a **pulsed** laser. A chopped laser throws away a lot of light. There are reasons to use a chopped laser beam and a common optical lab component is an optical chopper that looks basically like a fan blade that you put in the beam path.

A pulsed laser takes the photons that would have ended up being thrown away by the chopper and moves them into the laser pulse. You can think of a pulsed laser as a laser focused in time rather than in space. A pulsed laser gives very high

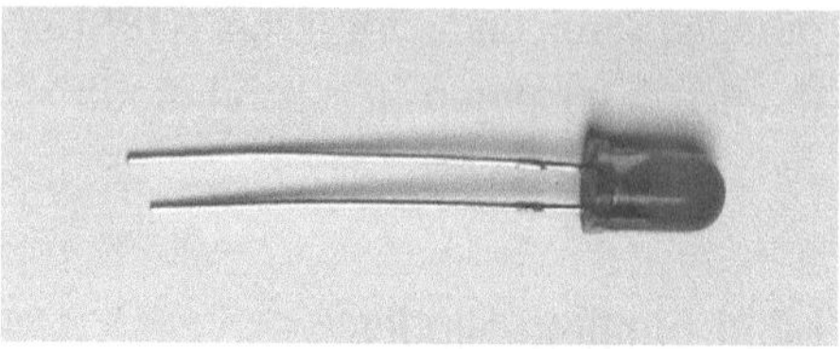

Figure 2.8

LEDs are polar devices, meaning it matters which way they are connected to an electrical source. The length of the leads is the key.

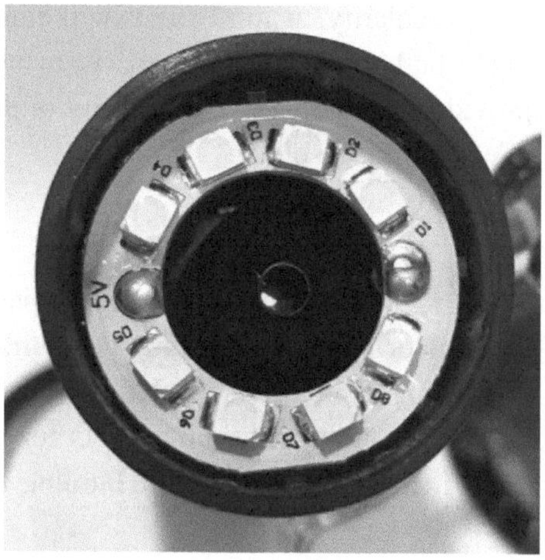

Figure 2.9

A ring of white light LEDs used for illumination on a digital microscope.

electric fields for a very short period of time, which can be used in a number of interesting ways, including nonlinear optics.

Pulsed lasers are characterized by their pulse length (how long the pulse is in time) and their repetition rate (how often the pulses are coming). The three most common regimes for pulse lengths are the nanosecond (10^{-9} s) regime, the picosecond (10^{-12} s) regime, and the femtosecond (10^{-15} s) regime.

Femtosecond lasers are commonly available now. In the femtosecond time regime, electrons in a sample can respond to the electric field of the laser pulse, but atomic nuclei cannot as they are simply too massive to respond that quickly. This allows for time-resolved measurements of the electron dynamics in a sample.

Another area where pulsed lasers are needed is in the field of **nonlinear optics**, involving phenomena that typically require higher laser power. In linear optics, you typically put in one photon and get out one photon. In nonlinear optics, multiple photons in may result in only one photon out. **Second harmonic generation (SHG)** is a classic example of nonlinear optics. In SHG you input two photons to a nonlinear material and get one back at twice the energy of the two individual photons that you put in. Nonlinear optical phenomena are used in imaging techniques collectively known as multiphoton microscopy. Examples of multiphoton microscopy are SHG imaging and two-photon fluorescence imaging (Sections 4.11.1 and 4.11.2).

Most pulsed lasers consist of solid state gain media which are pumped by a high-power (>5 W) DPSS, diode, or gas laser. Pulses are generated through two schemes known as Q-switching and mode locking. Q-switching is usually used for nanosecond lasers and mode locking for picosecond and femtosecond lasers.

Q-switches are special shutters that open only when a laser pulse has obtained the desired **quality factor** or **Q-factor**. A laser cavity is set up as described previously and often a pulse of light from a xenon flash lamp is used to pump the gain medium. A laser pulse starts to circulate in the cavity, gaining intensity with each pass through the cavity. When the pulse has been amplified enough, the Q-switch is opened and the pulse is sent to the output port of the laser. Q-switches often work with an electro optic crystal. Electro optic crystals will change from acting as a mirror to a specific polarization of light to being transparent when a high voltage is applied to the crystal. Q-switched lasers can give extremely high-pulse energies, but are often limited in their repetition rate and pulse duration by the Q-switch process. Q-switched lasers are typically in the nanosecond and picosecond pulse duration range. Repetition rates of 10 or 20 Hz are common for Q-switched laser systems.

Mode-locked lasers, usually titanium sapphire based systems, work by having a laser cavity that will support multiple modes (wavelengths) of lasing. These modes will constructively and destructively interfere, giving rise to some positions with no laser light and some positions with lots of laser light. When the modes are locked together the pulse train becomes stable typically around a 76 MHz repetition rate in the most common systems. Mode-locked lasers give femtosecond or picosecond pulses at lower overall pulse energies compared to Q-switched nanosecond laser systems.

Pulsed lasers often offer extremely high powers and highly collimated beams, but historically have been more complicated and take a bit more alignment and patience than CW lasers, which often just "work" when you plug them in and turn them on. Modern automation has made femtosecond lasers also plug and play systems now.

Pulsed lasers are normally named based on their gain media. Some common ones are Nd:YAG, Ti:sapphire (commonly pronounced as "tie-saff"), which

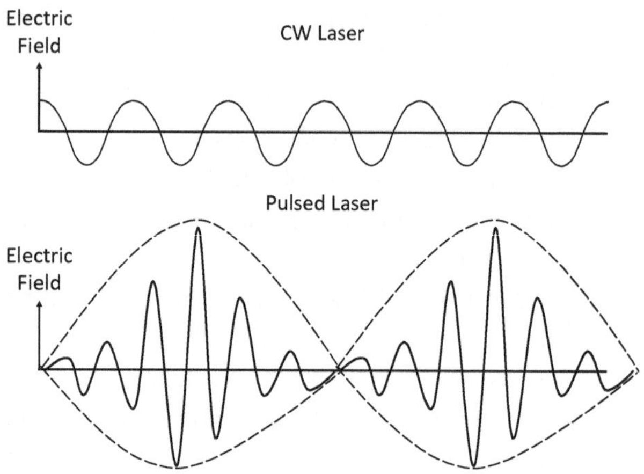

Figure 2.10

Comparison of the electric field of a CW laser and a pulsed laser. The dotted lines of the pulse laser are referred to as the pulse envelope.

consists of a sapphire crystal (Al_2O_3) doped with Ti ions, Nd:YVO (neodymium-doped yttrium orthovanadate), Nd:YLF (neodymium-doped yttrium lithium fluoride), and excimer lasers. Historically, ruby (essentially Cr:sapphire) was used as the first pulsed laser but the fastest pulses achieved were on the order of milliseconds.

The two most common pulsed lasers are the Nd:YAG and the Ti:sapphire. Nd:YAG lasers are typically in either nanosecond or picosecond systems. Nd: YAG gain media have a fundamental emission wavelength of 1064 nm that comes from **electronic transitions** in the neodymium ions spaced throughout the crystal. Nd: YAG lasers are often used frequency doubled to 532 nm.

Ti:sapphire lasers can yield either picosecond or femtosecond pulses. When someone says ultrafast laser or femtosecond laser, they are usually referring to a Ti:sapphire laser. Ti:sapphire lasers typically work in the range of 700–1000 nm. Some can be tuned to a specific wavelength. Erbium-doped fiber lasers are also an option in the femtosecond regime with a fundamental wavelength of 1030 nm.

2.2.6 Harmonic Units

Nonlinear crystals are most commonly used to frequency double or triple a laser beam. For instance, a Nd:YAG laser fundamentally produces light at 1064 nm. If you send the pulsed laser through a **second harmonic generation (SHG)** crystal, it will absorb two 1064 nm photons and emit one 532 nm photon. If you want shorter wavelengths, you can send the remaining 1064 nm beam and the new 532 nm beam through a third harmonic generation (THG) nonlinear crystal and it can absorb one 1064 nm photon and one 532 nm photon and emit a new 355 nm photon.

TIP: Harmonic generation is not a perfectly efficient process. You will often have a nontrivial amount of 1064 nm light in a 532 nm laser generated by an SHG process. If you are trying to do spectroscopy and are pumping at 532 nm, you will probably see a peak at 1064 nm that is not photoluminescence from your sample but is merely residual fundamental light.

2.2.7 Optical Parametric Oscillator

An **optical parametric oscillator (OPO)** is basically a resonant optical cavity that produces a laser beam with tunable wavelength. OPOs work by using a nonlinear crystal which absorbs one photon then re-emits two photons of different energies. The exact energy split between the two emitted photons depends on the angle of the nonlinear crystal to the electric field of the input light and the temperature of the crystal. Optical parametric oscillators can be angle tuned or temperature tuned. Angle tuning offers the clear advantage of being able to change wavelengths faster. The beam with the higher-energy photons is traditionally referred to as the "signal" beam, and the lower-energy photon beam is referred to as the "idler" beam. The signal and the idler beam are usually spatially separated as they exit the nonlinear crystal. An OPO needs to be pumped by a pulsed laser source. Many modern tunable femtosecond laser systems have an OPO under the hood to extend the spectral tuning range.

2.2.8 Optical Parametric Amplifier

An **optical parametric amplifier (OPA)** is similar to an OPO but with an amplification stage to add more energy to every laser pulse. The amplification stage usually has another lasing medium crystal that is being actively pumped as a source for the extra energy.

2.2.9 Characterizing Your Light

There are three things you often need to know about your light: what wavelength it is (what color); how much light there is (power); and, if it is a pulsed laser, the characteristics of the pulse (repetition rate and pulse width).

To measure the wavelength of light, you can use a **spectrometer**. A spectrometer consists of a diffraction grating which spreads out the incoming light like a rainbow. Nowadays a digital camera takes a picture of the rainbow and tells you how many photons of each color you have. Spectrometers, diffraction gratings, and digital cameras are all discussed in greater detail in the optical components section (Sections 2.3.36, 2.3.38, and 2.3.47).

To measure your laser beam's power (or the energy per pulse), you use a **power meter**, which consists of a sensor and a readout device to turn the recording from the sensor into a power. In newer models, the readout device can interface with a computer. Semiconductor photodiodes, thermopile, and pyroelectric are the most common types of power sensors. Each has a different working principle, so they are appropriate for different powers and wavelengths.

Be sure to find a power meter that is suitable for your power range and wavelength of interest. For instance, if you put a semiconductor diode meter meant to measure a few milliwatts in front of the pulsed laser we discussed above ($P_{avg} = 0.76\,W$), you will permanently damage your power meter. Another consideration is the noise level of the sensor – if you need to measure with nanowatt precision, be sure to purchase an optical sensor with extremely low noise. Working principles and the type of measurement of each type of power meter are summarized in Table 2.1.

Characterization of pulsed lasers requires a few basic formulas, since their power output is not constant with time, as shown in Figure 2.12.

Energy (in joules) is used to describe pulses (not CW light) and is defined as

$$E = \text{peak power} \times \text{time of laser pulse duration.}$$

In the lab, pulse durations can range from milliseconds to femtoseconds and energies per pulse can be as high as 100 mJ for longer (e.g., ms) pulses. The average power is

$$P_{\text{avg}} = E \times \text{frequency of pulses} = E \times 1/\text{period.}$$

The pulse duration is the period of time over which light is emitted. But, another important quantity that characterizes pulsed lasers is the frequency of pulses, which is known as the **repetition rate**, or "**rep rate**" for short. The rep rate is given in hertz (Hz; 1/s). Rep rates in the research lab typically range from 10 Hz to 80 MHz, and describe the number of short pulses that a laser emits per second.

Table 2.1 Types of power meters, working principle, and use/power range.

Power meter type	Working principle	Power/ energy range	Wavelength range	Appropriate measurements	Other
Semiconductor photodiode (optical)	Converts light to electricity	10 nW to 50 mW	280–1800 nm	Average power (CW only!)	Can only handle <10 s of mW – be careful!
		10 pJ to 800 nJ	280–1800 nm	Energy per pulse	Energy per pulse requires a special configuration
Thermopile	Converts light to heat	200 μW to >5 kW	0.15–12 μm	Average power (CW or pulsed)	Fairly slow; takes several seconds (or minutes for >kW sensors) to heat up
		1 mJ to >300 J	0.15–12 μm	Single long (>1 ms) pulse energy	Measurement of pulse energy requires special electronics
Pyroelectric	Converts light to a temporary voltage	100 nJ to >10 J	0.15–12 μm	Energy per pulse	Fast (<ms) response time

Note that the inverse of rep rates does *not* yield the pulse length – this is because the **duty cycle** (or the percent of time that the laser is actually emitting light) is <100 percent.

For instance, if your Ti:sapphire laser has a 76 MHz rep rate, 200 fs pulse duration, and 10 nJ per pulse, your peak output power is

$$P_{\text{peak}} = 10\ \text{nJ}/200\ \text{fs} = 50\ \text{kW}$$

and your average power is

$$P_{\text{avg}} = 10\ \text{nJ} \times 76\ \text{MHz} = 760\ \text{mW} = 0.76\ \text{W}.$$

This is a reasonable average output power for pulsed a Ti:sapphire laser. (Note that 0.76 W average power at 800 nm, the typical emission wavelength of a Ti: sapphire laser, is more than enough to burn your skin! Reaching into this beam path feels not unlike a bee sting.)

Figure 2.11

Two types of power meter heads. At the top is a thermopile detector which measures the power by converting photons to heat. Thermopile detectors are largely wavelength independent. At the bottom is a silicon-based detector which converts photons to electrons and counts the number of excited electrons.

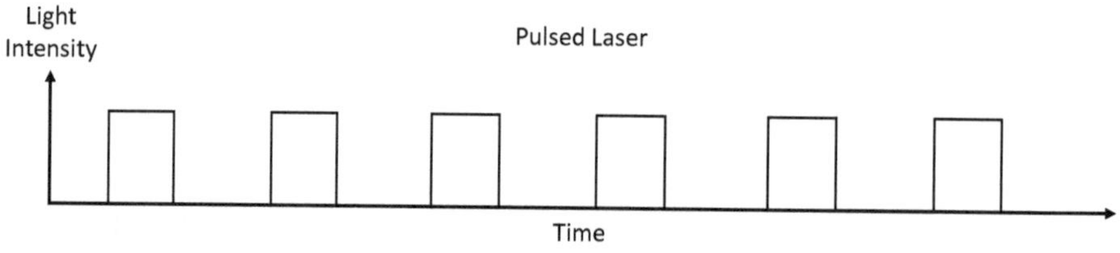

Figure 2.12

Light intensity is the square of the electric field amplitude. Pulsed laser intensities are often depicted with a square wave approximation even though they are really Gaussian (or $sech^2$) pulses in time.

2.2.10 Autocorrelators

To measure the pulse length of an ultrafast laser, you need to use some nonlinear optical tricks. First, you take the pulsed laser beam of interest and split it into two optical paths. One of these paths will have a fixed length. The other optical path will have a mirror that will scan back and forth, making the path length shorter and then longer than the fixed path. You then recombine the two pulses in a nonlinear crystal. The nonlinear crystal converts each individual pulse into an SHG beam at twice the frequency of the input beam. The second harmonic beam will be "**co-linear**" with the fundamental beams, meaning that they travel along the same optical path. The nonlinear crystal also yields a sum frequency generation (SFG) beam with the sum of the energies of the two input beams, which will be spatially separated from the two input beams. When the two optical path lengths are identical, you have the highest concentration of photons in time and space and therefore have the maximum sum frequency signal. Since the speed of light is constant, the change in distance of the scanning mirror is directly related to the

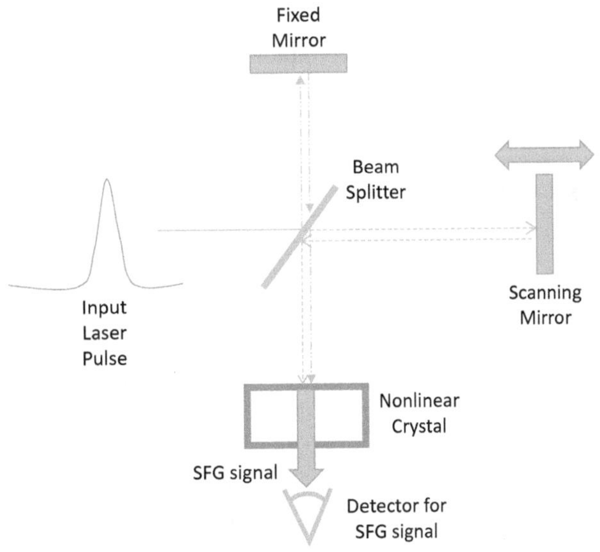

Figure 2.13

Schematic for an autocorrelator to measure ultrafast pulse lengths.

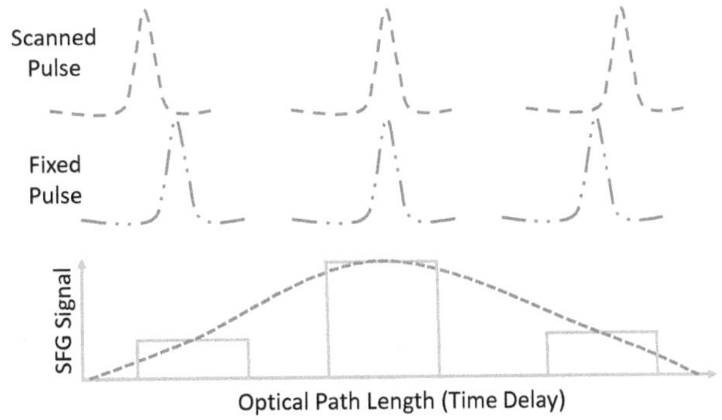

Figure 2.14

Detection scheme to determine pulse length with an autocorrelator.

time at which the pulse arrives at the nonlinear crystal. A 100 fs pulse is 30 μm long. (A typical human hair is ~100 μm in diameter.) By measuring the intensity of the SFG signal as a function of the position of the scan mirror, the physical length (which is directly related to the pulse length in time) of the pulses in the beam of interest can be determined.

2.3 Common Components in an Optics Lab

The first time you step into an optics lab, the array of specialized objects can be a little overwhelming. There tends to be a method to the madness and hopefully this section will help you understand how some of the common optical components fit together and are used in optical setups.

2.3.1 Optical Tables

The natural place to start a discussion on how to build an optical setup is with the **optical table** (aka **laser table**) on which you will build your setup. Typically, optical tables are covered by a very large, very heavy, and very flat metal slab (usually stainless steel) drilled with lots of screw holes in a square grid. Optical tables are designed to minimize vibrations that will mess up measurements and slowly throw your carefully constructed system out of alignment. An optical table has several engineered features to try to damp vibrations. The metal slab sits on a black box that contains a honeycomb-type structure that helps to damp vibrations. The whole assembly often sits on four or six vibration-damping legs containing a piston. The piston is filled with air and mechanically decouples the table from the floor. This allows the table to be isolated from vibrations caused by such things as the building elevators, trucks going by on the street, construction down the block, etc. These tables are often known as "**floating tables**" or "**air tables**."

TIP: Check that your optical table is actually floating! Many optics tables in labs are not floated properly. Gently press the edge of the table and the entire table should easily move down a little. When you remove your hand, you should hear a gentle hiss as the pistons in the legs refill with air and the table should return to its original level. If the table doesn't move when you press on it, your table is not floated. Fixing it is usually easy. As a first step, trace the airline back to the compressed air source, typically "house air," which usually has a valve and pressure gauge on the lab wall, and check that you actually have air pressure. Step two is to crawl under the table and check that there is a gap between the table baseplate and the solid part of the leg. Please watch your fingers around the gap. Optics tables are extremely heavy and can tilt quickly if someone unwittingly leans on them.

If there is no gap you will need to adjust the air pressure in the piston. There is usually a little threaded bolt that is used for a mechanical feedback loop to adjust the air pressure in the pistons. By adjusting the bolt so that it presses down on the lever to which it is attached, the air pressure in the piston will increase, lifting your table. The manufacturer's specification for the table will advise on the proper air pressure to use for floating the tables.

TIP: Keep a bubble level (available from a hardware store for <$10) around the lab. It is useful for many purposes, including making sure each leg of your optical table is equally inflated and the optical table is level.

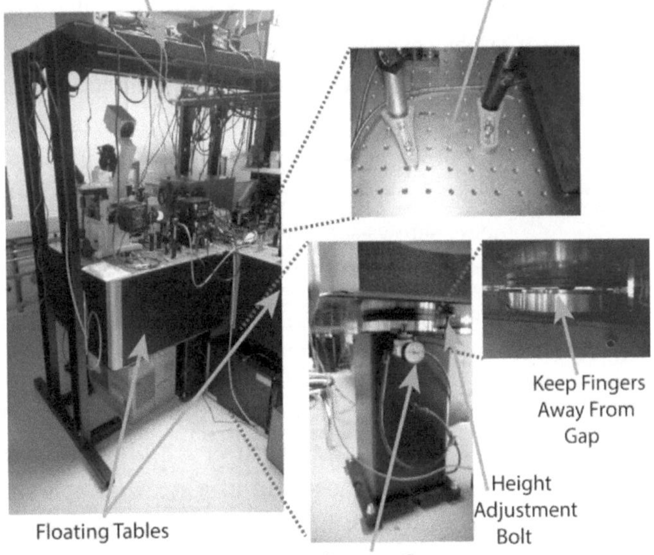

Figure 2.15

Floating optical table. In this case, we show two floating tables positioned at right angles adjacent to each other. For use of the adjustment bolt and details of how to float the table, please refer to the manufacturer's instructions.

Depending on the size of the table (which can range from several feet square to SUV-sized) and your requirements for the table to be floating or to be magnetic, the whole setup may cost from $2,000 to $20,000 new.

Optical tables are designed with either imperial (typical in the United States) or metric measurement standards. On imperial tables, the screw holes are spaced 1" apart and are "tapped" (meaning that the holes are threaded to accept screws) with **¼-20** (pronounced "**quarter-twenty**") holes. The phrase "¼-20" means that the screws are ¼" in diameter and have 20 threads per inch along the length of the screw. Typically **cap screws** (also known as **socket screws**) are used on an optics table. They usually have a hexagonal indentation, or socket at the top such that you tighten with a **hex key** (aka **Allen wrench**™). These types of screws can be purchased in many lengths from an optics company or from a hardware store for just a few dollars. Optical supply companies also sell very helpful kits with screws of many lengths and appropriate hex keys. On metric tables, the screws holes are spaced 25 mm apart and conform to the "**M6**" standard, meaning that they have a 6 mm diameter; corresponding screws and kits are readily available.

TIP: In the United States you will inevitably end up with a mix of metric and imperial screws in your lab because many instruments (lasers, microscopes, etc.) are manufactured outside of the United States and use metric screws. The convention (although it is not always followed) is for imperial screws to have a ribbed cap and a metric screw to have a smooth cap. A sure way to mess up a hole on the optics table is to force an M6 screw into an ¼-20 hole. Never force a screw into a hole. If it seems like it doesn't want to screw in, check that you have the right size screw. We suggest labeling your screw sets if they are not already labeled with either "imperial" or "metric." Try to keep imperial and metric screws separate so as to avoid destroying your components.

TIP: When using a screw to secure a large, unsteady, or oddly shaped optic, use a washer at the head of the screw. A washer is a metal ring that fits around the threaded part of a screw or bolt and sits snugly under the head. Using a washer helps to distribute the force applied by the screw and to stabilize the optic. You can see the washers on the left-most screws in Figure 2.16.

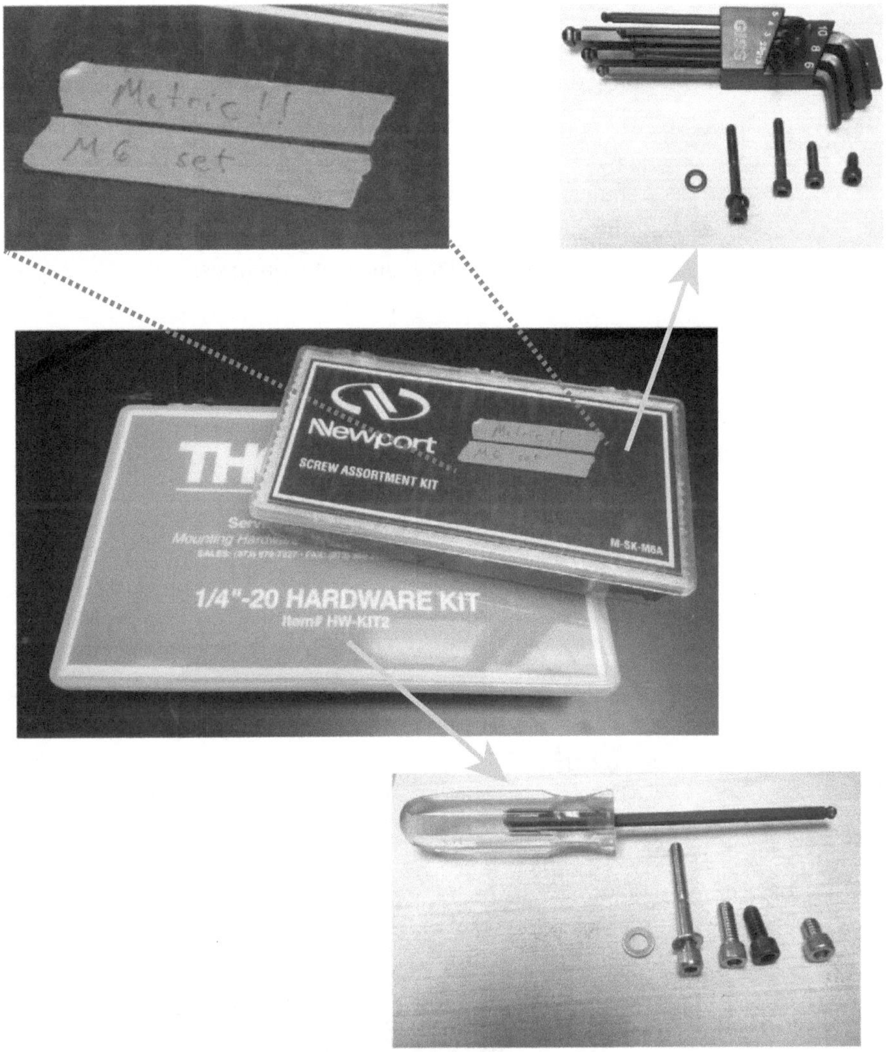

Figure 2.16

Imperial and metric screw sets. The image of imperial components (bottom) shows a hex key and several ¼-20 cap screws with compatible washers. The image of metric components (top right) shows several M6 screws and suitable hex keys for tightening them. As shown top left, always label your screw sets to keep imperial and metric organized and separate.

The other type of screw typically found in an optics lab is a **set screw**. Set screws look like threaded cylinders with a hexagonal socket in one end. In optics labs set screws are often used to hold two threaded components together. In the imperial system you will typically encounter ¼-20 and **8-32** (pronounced "**eight-thirty-two**") set screws.

TIP: Set screws, in particular (and to a lesser extent cap screws), frequently roll onto the floor from tables and desks. A plastic petri dish is a great way to keep all your screws corralled as you work. As we discussed in the safety section (Section 2.1), if your light source is on, don't bend over to pick up fallen screws – or anything on the floor! Either find another screw or turn off the laser before picking up the fallen screw. This is a common way for people to get laser induced eye injuries.

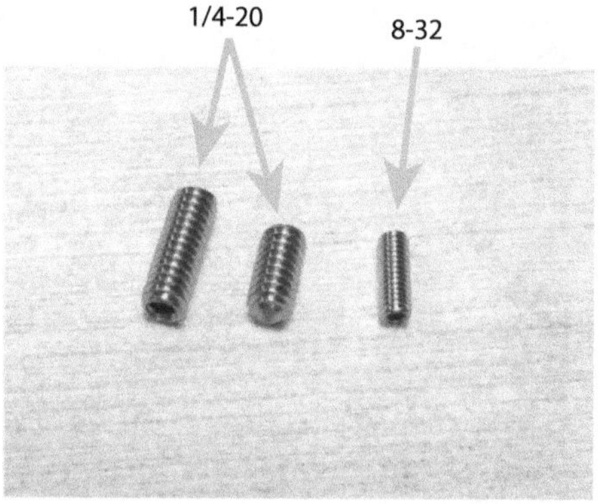

Figure 2.17

¼-20 and 8-32 set screws. Notice the hexagonal openings at one end of each set screw such that they can be screwed in or out with a hex key.

Figure 2.18

Screws in a petri dish. Small plastic dishes are terrific vessels to prevent small components, especially set screws, from rolling off the table and falling on the floor.

If you're less worried about vibrations, the cost of optical tables is too high, or your setup is not too large or needs to be portable/installed in another piece of equipment, you can purchase smaller alternatives to full laser tables. Referred to as an **optical breadboard**, these metal plates still have grids of screw holes but are small and light enough to lift yourself and position wherever you want. Other alternatives for vibration dampening are sometimes viable options depending on your requirements. "Sad balls" look like small black rubber balls, but they are made of a polymer (polynorbornene) that absorbs energy easily. If you try to bounce a sad ball on the floor it will "thud" and absorb all the energy. They are available from many sources on the internet. Sand bags or even polystyrene are possibilities if you need some vibration damping in a field environment.

TIP: After all the effort to damp vibrations, the worst thing you can do is place anything with a cooling fan (computers, laser power supplies, amplifiers, water chillers or circulators, etc.), motor, or pump on the table. These are sources of vibrations. Often, optics labs have shelves built over the optics tables that are suspended from the ceiling to mechanically isolate all of these vibration sources from the optical table, but still have the systems close enough to the table to be usable. Just be careful not to hit your head on these shelves. You can add extra padding (e.g., pipe insulation foam and duct tape) at particularly troublesome corners in order to protect your head.

2.3.2 Optomechanics

Optomechanical components are very modular and can be easily reconfigured to build different setups. Having a good understanding about how these components can be assembled will make building an optical setup much easier. Let's start at the surface of the optics table and work our way up. As mentioned in the previous section, the optics table has a grid of threaded holes (either ¼-20 or M6) to allow you to secure optical elements in various positions. There are a couple different systems of **base plates** for anchoring post holders to the optics table. One popular system is to use post holders with metal flanges at their base. Some post holders have the flanges permanently fixed. Other post holders have small metal plates that screw into the bottom of the post holder to make this flange base. The flanges allow for **clamping forks** to secure the post holder to the table. The clamping fork has a two-pronged "fork" that will fit over the baseplate flange and a ¼" wide slot to accommodate various positions for the ¼-20 cap screw that you will use to secure the clamping fork to the optics table. Generally, clamping forks allow for good versatility during alignment and excellent mechanical stability. The other common system for securing the post holders to the optics table is to screw a metal base with a ¼" slot for the cap screw directly into the bottom of the post holder. These baseplates come in a couple different shapes (see Figures 2.23 and 2.24). Please note that the base plates have a countersunk through-hole in them so they can rest flush with the table when the base is bolted to a post holder with a ¼-20 screw. A **countersunk through-hole** means that the hole has a wider portion to accommodate the cap of the cap screw and that the hole is smooth (not tapped) on the inside.

2.3.3 Post Holders

Post holders are basically hollow tubes of various lengths that posts slide into. The combination of a post and a post holder allows you to adjust the height of the optical element mounted on top of the post. **Thumb screws** in the sides of the post holders tighten to hold the post at a specific height. Please purchase locking thumb screws. Locking thumb screws can be tightened with an Allen wrench and provide much greater stability than the spring-loaded thumb screws that can only be tightened "finger tight." Many swear words have been uttered over loose thumb screws destroying a carefully aligned setup.

Figure 2.19

Left: locking thumb screw that can be tightened with a hex wrench. Right: a spring-loaded thumb screw that can only be tightened finger tight.

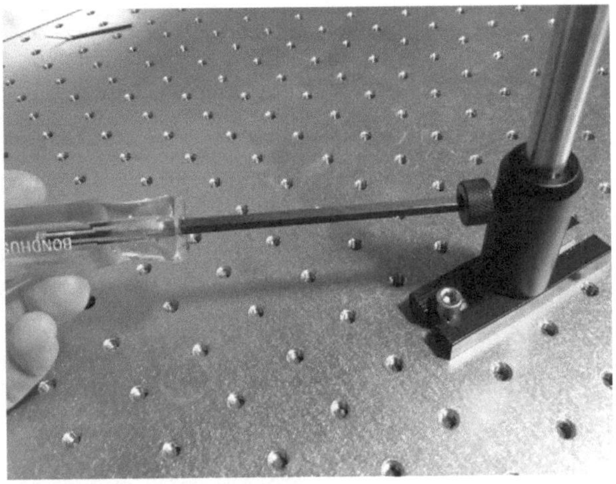

Figure 2.20

Locking thumb screws allow you to tighten a post into the post holder with a hex wrench, making your optics setup more likely to stay aligned.

TIP: It is good practice to set up your optical paths in straight lines. Using the grid of holes on the optics table as a guide works well. Many optical components are designed to work at either 0 degree angle of incidence or 45 degree angle of incidence.

2.3.4 Posts

Posts are metal rods that are used to control the height of the optical components, such as lenses and mirrors that guide the light around the optical table. In the imperial system they are ½" in diameter, with threaded holes at both ends. One end has a ¼-20 thread and the other has an 8-32 thread. Posts come in various lengths (from 0.5" to 10") and can be purchased with ruled distances on them to give you somewhat more precise height adjustment if you need it. If needed, you can combine two short posts to make a long posts with a set screw, although a solid single-piece post is much more stable in the long run than two that are screwed together.

Posts do come in 1" and 1.5" diameters, but these are typically used for setups which need extra resistance to vibration/drift or to support heavier objects. Larger diameter posts also have ¼-20 and/or 8-32 tapped holes for combining them

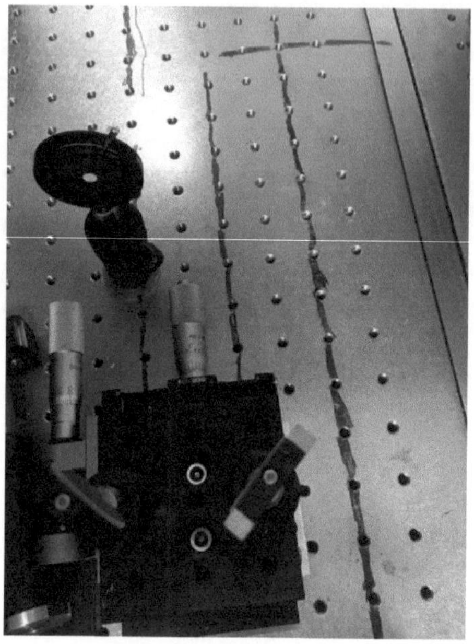

Figure 2.21

Drawing the beam path directly on the optics helps everyone remember the optical layout and where the laser beam is going. Laying out the optical path to go along the line of screw holes in the optics table helps make sure that the beam in your setup is going in a straight line.

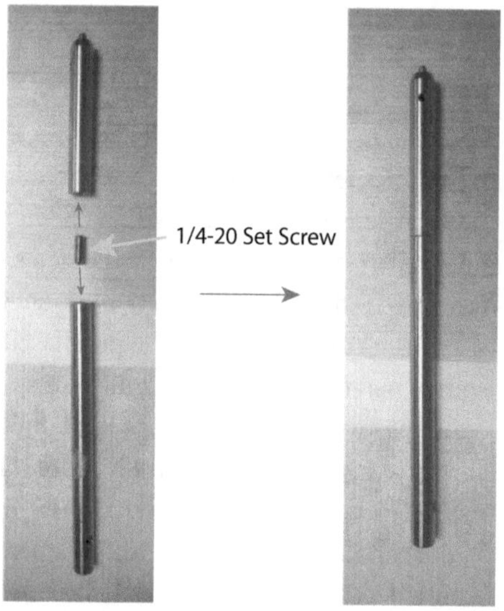

Figure 2.22

Combining two ½" diameter posts into one longer post using a ¼-20 set screw.

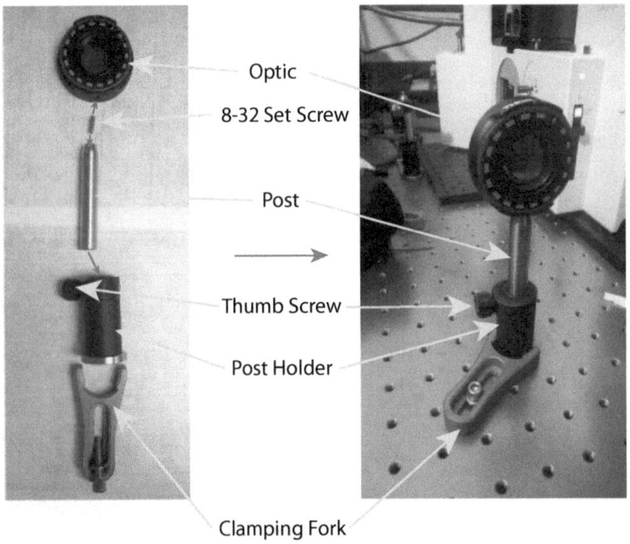

Figure 2.23

Assembled optical mount using a clamping fork base. A washer is used between the screw and the clamping fork for extra stability. In this case, the post holder has a flange at its base, allowing the clamping fork to clamp it into position. Note that use of a clamping fork allows the optic to be positioned anywhere within a certain distance (determined by the length of the clamping fork) of the screw.

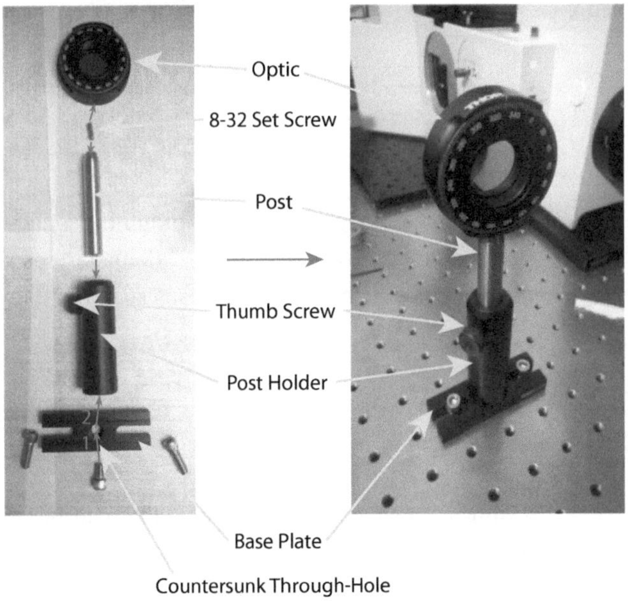

Figure 2.24

Assembled optical mount using a flat base plate. In contrast to post holders used with clamping forks, post holders for flat base plates do not have a flange at the bottom. In this case, an additional ¼-20 screw is used to attach the post holder to the base plate through the countersunk through-hole in the base plate.

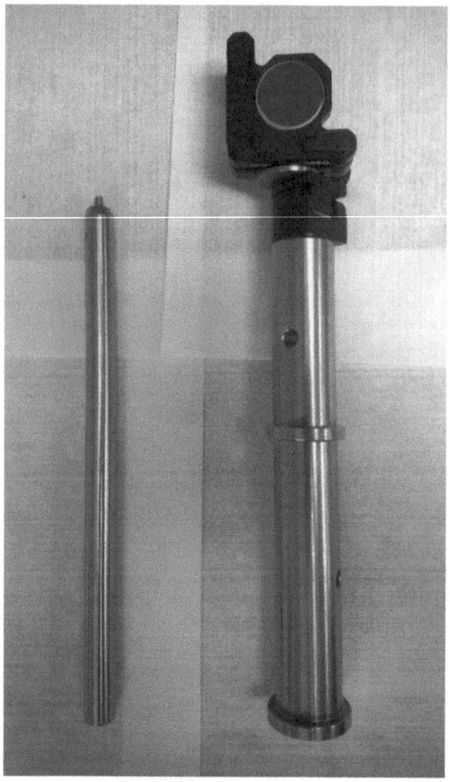

Figure 2.25

Comparison of a ½" and 1" diameter post. In this case, the 1" diameter post has an aluminum mirror mounted on top of it.

together into longer posts and mounting optics. Naturally, 1" or 1.5" diameter posts are more expensive than ½" diameter posts. Many 1" or 1.5" diameter posts also can have flanges at their bases such that they can be positioned and mounted with a clamping fork (Figure 2.23).

2.3.5 Collars

Collars are small metal crescents that can be tightened around a post just above the post holder. When the collar is tightened on the post, the thumb screw in the post holder can be loosened. This allows you to rotate the post for coarse alignment without risk of changing the height of the optical element. It also allows you to remove an optical element and replace it at exactly the same height. They are very useful.

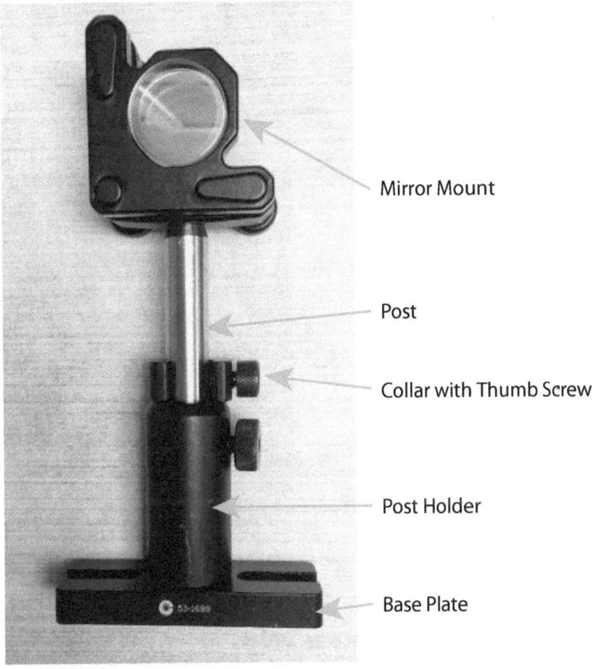

Figure 2.26

Collar installed on a post. In this case, the post is supporting a dielectric mirror (Section 2.3.13).

2.3.6 Post Clamps

Post clamps have two holes with thumb screws to slide posts into and will allow you to position a second post perpendicularly or at 45 degrees to the direction of the original post. As with post holders, we recommend post clamps with locking thumb screws. Post clamps can also be purchased for 1" or 1.5" diameter posts to mount optics at variable heights along the length of the post (see Figure 2.27).

2.3.7 Rails

Rails are dovetailed tracks that bolt to the optical table through several down-facing countersunk through-holes. The post holders are then mounted on bases (which have upward-facing countersunk through-holes) that can slide along the rail and can be locked in place on the rail with a thumb screw. They provide an alternative for systems with many optical components assembled in a line whose position relative to each other along the **beam path** may need to be adjusted. Rails

Figure 2.27

A right-angle post clamp used to mount a beam block perpendicularly to the axis of another post.

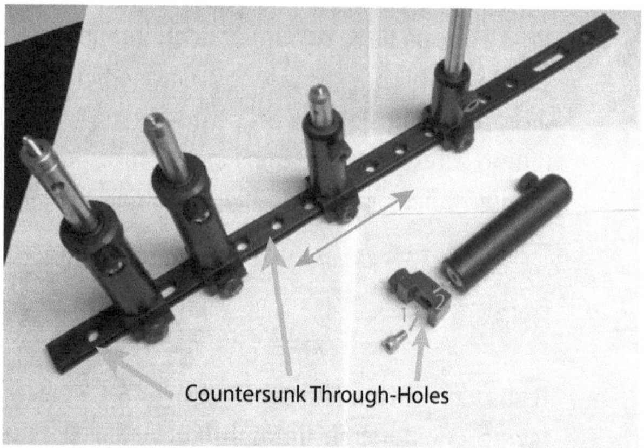

Figure 2.28

A rail system. By loosening the thumb screws on each base mount, components can be translated along the length of the dovetailed rail track. Rails are mounted to the laser table via countersunk through-holes; likewise, post holders are secured to rail bases in a similar manner.

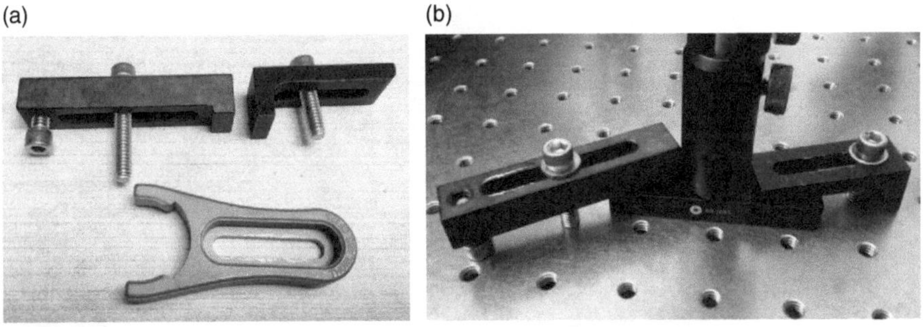

Figure 2.29

Table clamps. (a) Isolated table clamps. (b) A dogleg clamp and an L-bracket used to provide extra stability to a base plate.

allow you to slide components back and forth in the beam path without changing their left/right alignment or height (see Figure 2.28).

2.3.8 L-Brackets or Table Clamps

L-brackets and **table clamps** are used to mount optics which do not necessarily have screw holes or slots that match those on the laser table, or to provide extra mechanical stability to a large component. We suggest using a washer between the cap screw and the clamp to provide extra stability. Please note the orientation of the L-bracket and table clamp in Figure 2.29. The ¼-20 screw with a washer should be pushing down on a clamp or bracket that is parallel to the optical table. The table clamps and L-brackets are often misused, leading to less stable setups. If your bracket or clamp is not parallel to the table you probably have it flipped around backward. The table clamp to the left in Figure 2.29 has a threaded hole in one end. To use this clamp properly insert a ¼-20 screw into the threaded hole and adjust the height to the thickness of the object that you wish to clamp. The goal is to have a horizontal clamp in the end setup.

2.3.9 Removable/Flip/Kinematic Holders

Component holders which can easily move optical components into or out of the optical path are useful accessories to your optical toolkits. Often, these "**kinematics**" are spring-loaded or have magnetic alignment, allowing you to quickly flip, pop, snap or slide components into and out of the beam path without having to realign them. These are useful in countless applications. For instance, say you need to align your laser spot on your camera but don't want to image it

Figure 2.30

Demonstration of a flip mount to quickly and reversibly move a mirror into and out of the beam path.

when you collect actual data – you can use a flip holder to rapidly insert a filter to cut the excitation laser light. When the flip holder is down, the laser will reach your camera for alignment but when it's up, it blocks the laser, allowing you to collect your data. Likewise, magnetically clipped mirrors and bases are useful for switching back and forth from one laser to another or from your laser to a white light source. Flip and magnetic mounts and bases cost ~$100, though if you want to splurge you can buy computer-controlled flip mounts.

2.3.10 Micromanipulators

Sometimes you'll need to be able to move components around for alignment or adjusting the position of your sample or laser beam. To do so, you need micromanipulators, which enable sample or component movement by turning a knob or powering a motor. Newer electronic micromanipulators and stages can be computer-controlled, which is often desirable for scanning measurements, and can reach smaller step sizes/resolutions than by-hand controllers, though for many applications manual micromanipulators are sufficient. Micromanipulators can cost from several hundreds of dollars for a single-axis manipulator to several tens of thousands for a stable, computer-controlled, three-axis

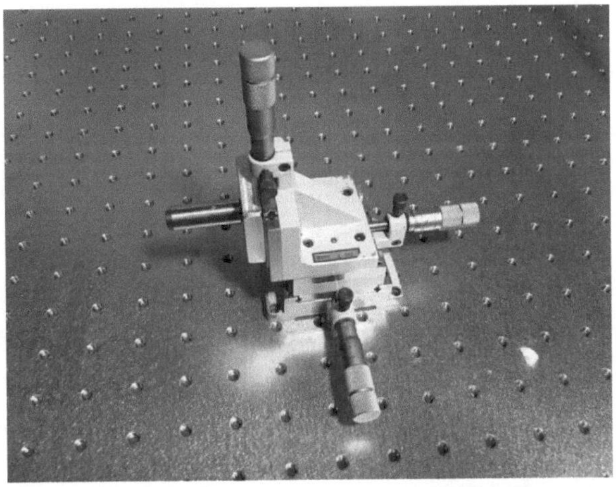

Figure 2.31

Three-axis manual translations stage.

manipulator with 100 nm step sizes. **Piezoelectric stepper motors** enable tiny step sizes of <50 nm if you're willing to pay for them.

Often, you can combine several single-axis manipulators into a two- or three-axis manipulator.

2.3.11 Cube/Cage Systems

Cage or **cube** systems are alternatives to rails or table-based alignment. They allow easy alignment using metal rods which screw into square or rectangular boxes and frames at various set positions, meaning that all components are aligned and centered simply by virtue of being in the inflexible cube network (Figure 2.32). Thus, cage systems are best for tricky alignments which require centering small features (like pinholes or apertures – Section 2.3.19) and for building complex, 3D "up and over"-type beam paths since the cube network provides the mechanical support required to mount optics far above your table. Cage systems come in three sizes: 16, 30, and 60 mm, which are ideal for ½-, 1-, and 2" optics, respectively. Cage-based mounts for most standard-sized components can be purchased. Some mounts have tapped screw holes for the ends of the metal rods, meaning their position can only be adjusted by changing the entire structure, whereas other mounts have non-tapped holes and thus slide along the metal rods, allowing their position to be easily changed. Set screws tighten to keep the sliding mounts in place. Some optics, such as cubic beam splitters, are often sold in cubic mounts for

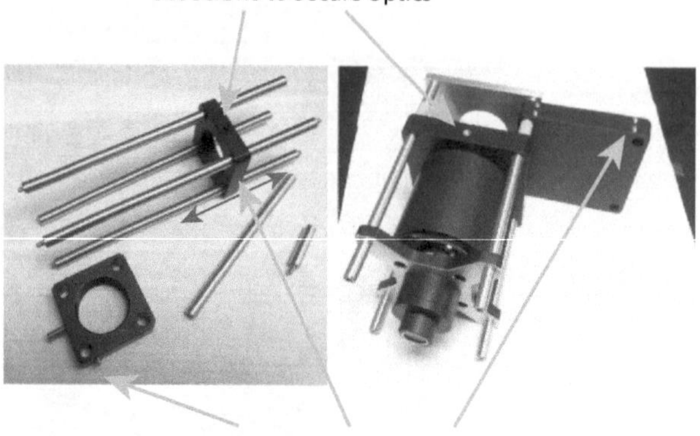

Figure 2.32

Cube systems. Left, a 30 mm cube system used for mounting 1" optics. Several extra posts and an optical mount are shown. The smaller set screws in the optical mount are used to lock the position of the mount along the post system; loosen these to slide the optic along. Right, a 60 mm cube system used to mount 2" components. Note that there are several optics mounted in series in the 60 mm cube system; cube systems such as these are ideal for mounting optics which need to be centered accurately with each other and/or for supporting heavy optics in a vertical configuration.

Figure 2.33

Kinematic cube mount. Magnets in the two sections of the kinematic cube allow rapid insertion and removal of a component at a certain angle into the beam path. This might be used to rapidly switch between sending light to detectors or further beam paths positioned at right angles to each other (e.g., switching between imaging and spectroscopy). In this case, a dichroic mirror is mounted at 45 degrees with respect to the faces of the cube system, but beam splitters, mirrors, and filters can also be mounted and rapidly inserted/removed in this way.

convenience, though it's not necessary to use them in a cage system since these mounts usually can attach to other posts with standard-sized set screws. Cube mounts can also be kinematic, as shown in Figure 2.33.

2.3.12 Optical Mounts

Optical mounts refer to components that hold your optical element to the top of a post (or, as discussed in Section 2.3.11, in a cube or cage system). Imperial system posts typically have threaded holes in each end. One end has a ¼-20 threaded hole; the other end has an 8-32 threaded hole. This 8-32 threaded hole is what is usually used to attach your optical mount to the post.

Various components require various optical mounts. We will discuss holders for each specific component in the later sections. Most optical components in the beam path will be circular with standard diameters. In the United States the standard diameters are ½", 1", and 2". Mounts for such circular optics are available with a variety of functionalities, and most of them attach to posts with 8-32 screws. In some cases, the mounts include their own micromanipulators or rotation stages for fine position manipulation and they might also fit into cube or cage systems. As discussed above, **kinematic mounts** imply some sort of motion as opposed to a **fixed mount**, which won't let you do fine position adjustments. Kinematic mounts also refer to mounts that allow you to rapidly insert or remove particular components from the beam path. We will discuss recommendations and best practice for optical mounts for each component in its respective section below.

Generally, your optical component will fit neatly into a ½", 1", or 2" circular opening in the mount. If the opening is threaded, a **retaining ring** can be tightened down with a **spanner wrench** to secure the optic (see e.g., Figure 2.47). If the opening is not threaded, the mount will have a set screw which can be tightened to secure the component (see e.g., Figure 2.36).

TIP: In all cases, we highly recommend wearing disposable, powder-free nitrile or latex gloves while handling all of your optics. Dust and fingerprints on your optics will scatter light and/or potentially absorb light and heat up, causing damage to your component. Only grasp the edges of your components; this way, even if dust does accumulate on your gloves for some reason (e.g., you accidentally brush your hair or your sweater), you don't scratch or contaminate the surface of the optic.

2.3.13 Mirrors

Just like the mirror in your bathroom, mirrors on optical tables reflect light. They are essential for steering the light from your light source to a location where you need it. As with so much, the story is just slightly more complicated when you start getting into the details – mirrors can be made of different materials which each present distinct advantages and disadvantages.

Metal mirrors are typically very smooth layers of gold (Au), silver (Ag), or aluminum (Al) with a transparent protective coating on top to protect the metal. Each metal has a specific wavelength regime over which they reflect well. Gold is excellent for working in the IR region with 97–99 percent reflectivity across the mid-IR region, but gold has very poor reflectivity below wavelengths of about 550 nm. Silver is typically the mirror of choice in the visible region (wavelengths of 400–700 nm) with 99 percent reflectivity in the visible region. UV-enhanced aluminum works in the visible and ultraviolet ranges down to a wavelength of 180 nm. One word of warning: although aluminum has a relatively flat reflectivity across the visible and ultraviolet range, it is only 90 percent reflective, so significant power losses can accumulate if you have a number of Al mirrors in your system. Metal mirrors cost $50–100. Because mirrors are so common, you can often purchase packs of ten at a somewhat significant discount. Far and away the most common size of mirror is the 1" (25.4 mm) diameter.

Mirrors can also be made from a stack of many layers of dielectric materials. Reflectivity and transmission of light off of or through stacks of thin films can be tuned by adjusting the thicknesses and materials that comprise the layers. **Dielectric mirrors** take advantage of this phenomenon and thus achieve extremely high (>99 percent) reflectivity, but generally over narrower wavelength

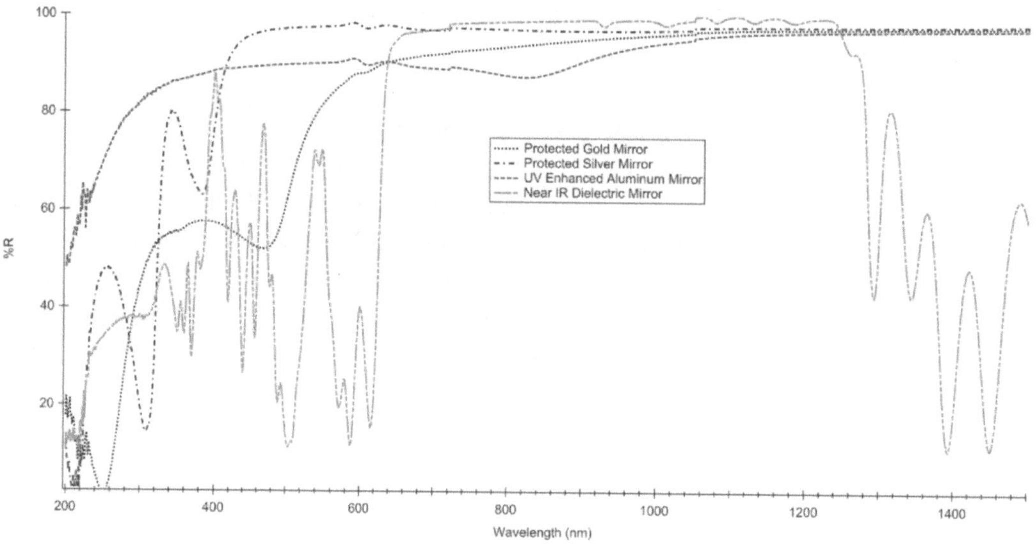

Figure 2.34

Reflectivity versus wavelength plots of three common mirror materials: silver, gold, and a broadband dielectric stack mirror designed for the near infrared region.

Figure 2.35

Top left: gold mirror; top right: silver mirror; bottom left: dielectric mirror designed to reflect 325 nm; bottom right: dielectric mirror designed to reflect in the near infrared 800–1100 nm.

regimes than metal mirrors. Dielectric mirrors also tend to have a higher damage threshold than metal mirrors, which is one of the reasons they are the mirror of choice in pulsed laser applications.

TIP: A single mirror will only allow you to change the direction of the laser beam. When setting up a beam path, always use at least two mirrors as this will allow you to translate the beam position, while maintaining the beam's direction. See "Walking the beam" Section 5.2.

Mirrors are typically mounted in spring-loaded kinematic mounts to allow for fine adjustments. The mount consists of two plates. The mirror is attached to the front plate, typically by securing the mirror in a countersunk hole with a small Teflon-tipped set screw that pushes against the side of the mirror, holding it in place. The front plate is connected to the back plate with a spring. One corner acts as a pivot point. There can be a small ball bearing or an adjustable screw to act as the pivot point. The other two screws on the mount are used to work against the spring and pivot the front plate in either the vertical or horizontal directions. Small

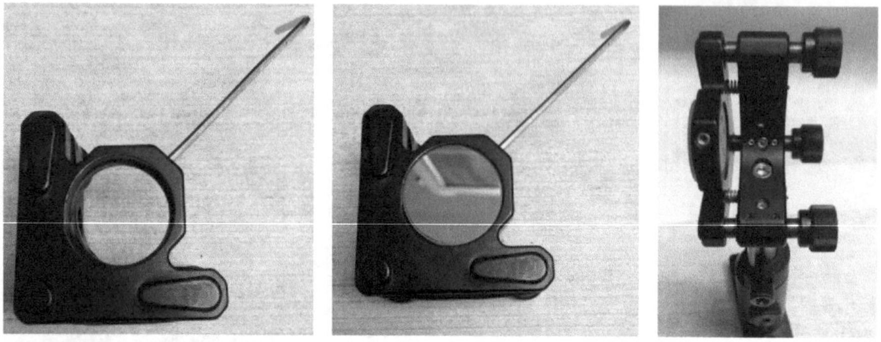

Figure 2.36

Mounting a mirror. An Allen wrench is used to tighten a set screw to keep the mirror in place.

adjustments in the horizontal and vertical tilt of the mirror allow you to steer the light around the optics table (see Figure 2.36).

TIP: When setting up the beam path, note that the front plate when the mirror is mounted on the kinematic mount is offset from the vertical line of the post. The plane that you care about for alignment purposes is the optical surface of the mirror. The vertical line of the post should be set back a little from the line of holes in the optics table in this case, so that the plane of the mirror is directly above the line of holes. Dielectric mirrors are particularly sensitive to the angle of incidence.

Most mirrors have a flat surface such that beam shape and size in are equal to the beam shape and size out. The roughness of the mirror can be described by λ/N, where λ is the wavelength of light that the mirror was designed for and N is an integer. The smoother the mirror, the less distortion your beam will suffer after reflecting off the mirror. Most mirrors feature a $\lambda/10$ roughness, which is suitable for many applications.

In specialized applications, you might encounter mirrors with convex or concave parabolic or spherical curvature to focus or disperse beams, just like lenses. The equations that govern focal length, magnification, and image/object size and position also apply to curved mirrors. However, unlike lenses, one advantage of curved mirrors is that they don't introduce chromatic aberrations. Concave mirrors are often used in light-collecting applications like telescopes or for focusing light in the UV or IR regions where materials with the proper optical properties to make a lens are harder to come by or manufacture. Convex mirrors are used to allow you to see around corners in hallways and parking lots and in the side-view mirrors of cars (hence the warning "objects in mirror are closer than they appear").

2.3.14 Polarizers

Polarizers selectively allow a single orientation of the electric field of light to pass. See Section 1.3 for an introduction to the concepts of polarized light.

Wire grid polarizers are the simplest polarizers. They consist of an array of parallel metal wires. When the electric field is parallel to the metal wires, the electrons in the metal can move along the wire to set up an equal and opposite electric field and the light is blocked from passing. When the electric field of the light is perpendicular to the metal wires, the electrons in the metal can't move far enough to set up an opposing electric field and the light can pass through the polarizer.

Remember that the orientation of the electric field of linearly polarized light can be described with vectors. A vector can be described by its x and y components. The polarizer will only block one component of the electric field vector. The practical effect of this is when the electric field of linearly polarized light is oriented 45 degrees to the orientation of the wire array, half of the light is blocked and half is passed. The wire grid blocks the vector component that is parallel to the wires and passes the vector component that is perpendicular to the wires. The light that is transmitted through the polarizer always has the same orientation relative to the polarizer, but the intensity of the transmitted light will depend on how large the vector component that is passed is. Thus a polarizer can be used to adjust the power of the laser, if the absolute polarization is not essential to your experiment. If you need to keep the polarization fixed, but still need to adjust the power,

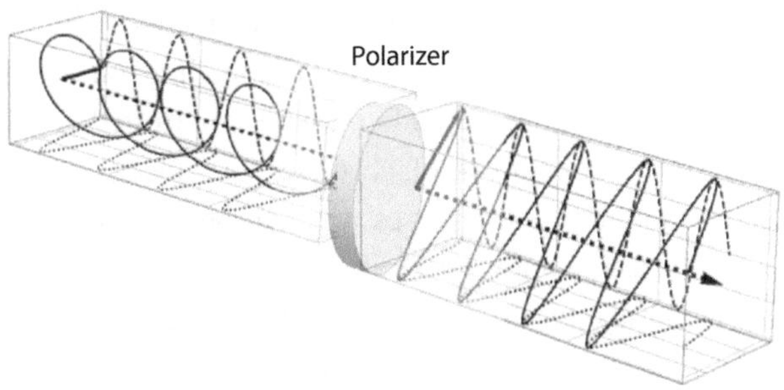

Figure 2.37

A circularly polarized beam passing through a polarizer will become a linearly polarized beam. The intensity of the linearly polarized beam will be one-half the intensity of the circularly polarized beam as the *x* component of the electric field has been rejected by the polarizer.

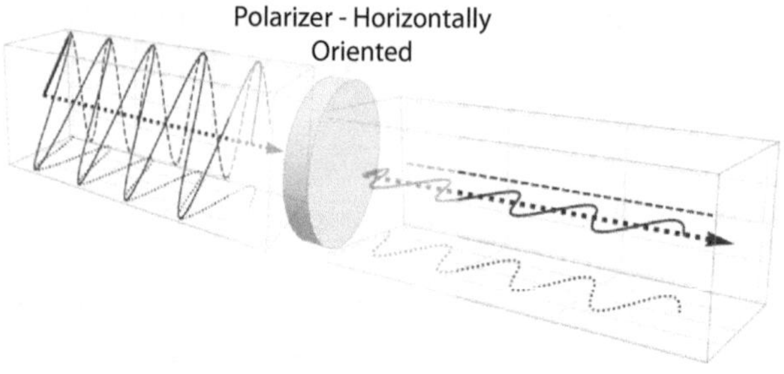

Figure 2.38

Linearly polarized light at 45 degrees to the polarizer results in linearly polarized light with one-half the intensity of the original beam oriented along the direction of the polarizer.

a half-wave plate (see Section 2.3.15) in front of the polarizer will rotate the orientation of the light electric field relative to the polarizer.

Wire grid polarizers are commonly used in microwave work, because the relatively long wavelength of microwaves (λ ~ millimeters to meters) means that it is easy to fabricate a wire grid polarizer with wire spacing on the order of the wavelength of the radiation. Nano- and microfabrication techniques have improved enough recently that wire grid polarizers in the visible light range are now available too. More commonly used, such as in your sunglasses, are long-chain conductive molecules that can be deposited in an oriented manner. The operating principle is the same as wire grid polarizers.

For applications that require higher extinction coefficients (how well the polarizer blocks the unwanted light), calcite polarizers are used. They work by using the birefringence of the calcite crystal to select only a certain polarization of light to pass.

Since the action of polarizing optics depends on the angle at which it is oriented with respect to the electric field of the incident light beam, mount polarizers in **rotation mounts**, which allow rotation of whichever optic is mounted in them. Rotation mounts that can be adjusted by hand can be purchased for all standard optic sizes for ~\$60–300, depending on the accuracy of angle positioning you need; electronic and computer-controlled rotation mounts are also available with varying degrees of accuracy and speed for ~\$700–2000. Polarizers come in all standard optical sizes and can be purchased to be compatible with many wavelength regimes. In all cases, polarizers typically have marks on their frame indicating the orientation of the transmission axis. The degree of extinction is a measure of how good a polarizer is. It is measured by taking two polarizers and putting them at 90 degrees to each other and measuring how much light is blocked. In theory, no light should get through crossed polarizers, but in reality some light always passes through the two orthogonal polarizers. Overall, what does get through is many orders of magnitude lower in intensity.

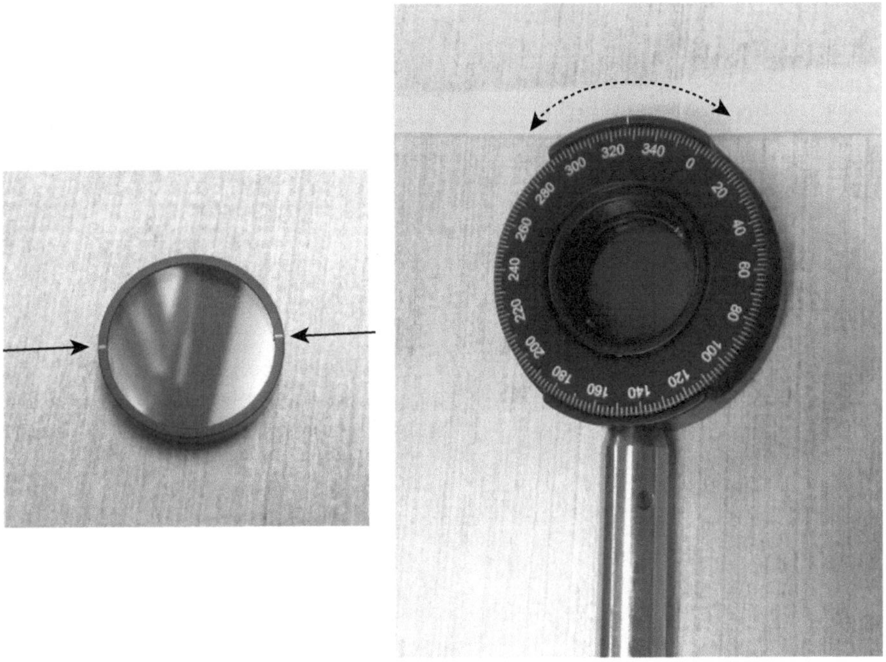

Figure 2.39

Linear polarizers. Left, an unmounted linear polarizer designed for the visible wavelength (400–700 nm) regime. Note the white marks on the frame indicating the orientation of the transmission axis. Right, a polarizer mounted in a manual rotation mount.

2.3.15 Waveplates

Waveplates adjust the polarization of light. **Half-wave plates** rotate the polarization of **linearly polarized light**, whereas **quarter-wave plates** convert linearly polarized light into **circularly polarized light** (or circularly polarized light into linearly polarized light). Waveplates are made of **birefringent** materials, meaning that polarized light passes more easily through the material at one orientation than another. This birefringence establishes a "**fast axis**" and a "**slow axis.**" If the incident polarization of linearly polarized light is aligned parallel with the fast axis of either a half- or quarter-wave plate, it remains linearly polarized in the same direction. However, for a half-wave plate, if the incident polarization is at an angle to the fast axis, the waveplate will rotate that polarization by twice the angle between the incident polarization and the fast axis. A quarter-wave plate will convert linearly polarized light to elliptically polarized light with its highest intensity at twice the angle between the incident angle and the fast axis. If the incident angle between the linearly polarized light and the fast axis of a quarter-wave plate is

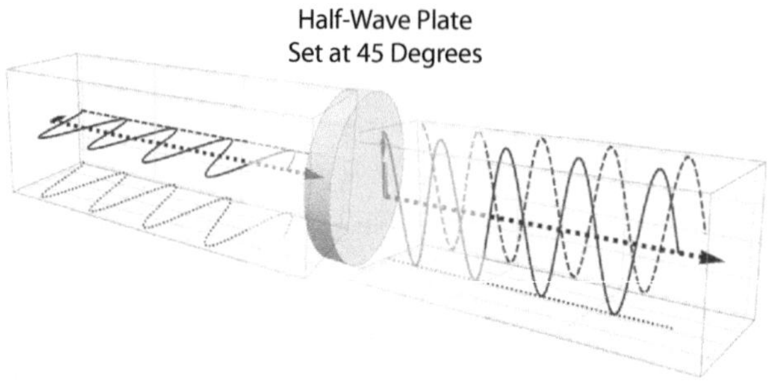

Figure 2.40

A half-wave plate (HWP or $\lambda/2$) can rotate the polarization of linearly polarized light without changing the intensity of the beam. The exact orientation of the electric field in the output beam depends on the orientation of the fast axis of the half-wave plate with respect to the electric field in the input beam.

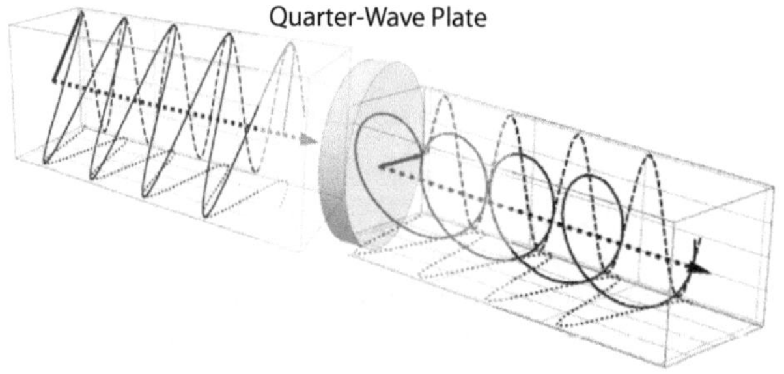

Figure 2.41

A quarter-wave plate (QWP or $\lambda/4$) will turn linearly polarized light into circularly polarized light.

exactly 45 degrees, the light emitted will be **circularly polarized**. The sign of the incident linear polarization (±45 degrees) determines whether the emitted circular polarization will be right-hand circularly polarized or left-hand circularly polarized.

Like polarizers, the angle of waveplates with respect to the electric field of the incident beam determines the nature of the exiting beam. So, waveplates should also be mounted in a rotation mount.

TIP: Be careful when you purchase your waveplates – since they depend on birefringence to work, they are designed for specific wavelength ranges. Furthermore, the **order** of waveplates determines their quality, with low order waveplates better than higher order plates.

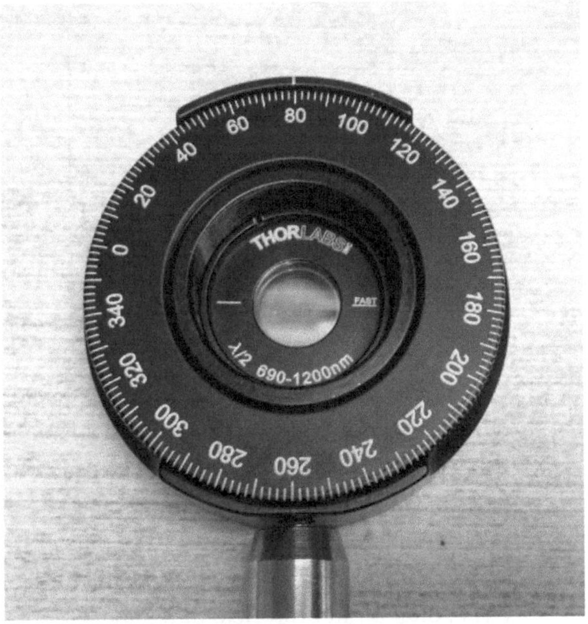

Figure 2.42

Half-wave plate mounted in a manual rotation mount.

2.3.16 Polarization Scramblers

To remove any polarization from your light beam, use a **polarization scrambler**. You might need to scramble your light's polarization if your spectrometer's diffraction grating reflects two polarizations of light with different efficiencies or to prevent any bias in your signal if your sample's emission or absorption might be dependent on polarization (as in the case of high aspect ratio structure like a wire or linear waveguide). Scramblers work by rapidly varying the polarization of a light beam in random directions, effectively eliminating any preference for a polarization direction with time. These devices can range from very simple depolarizing optics (~$600) to complicated electronic devices (>$10k), which are especially important for eliminating artifacts in fiber optic communication.

2.3.17 Beam Splitters

Beam splitters, as their name suggests, split one laser beam into two. They typically divide one input beam into two output beams at 90 degree angles (see Figure 2.43). You can purchase beam splitters with different **antireflection**

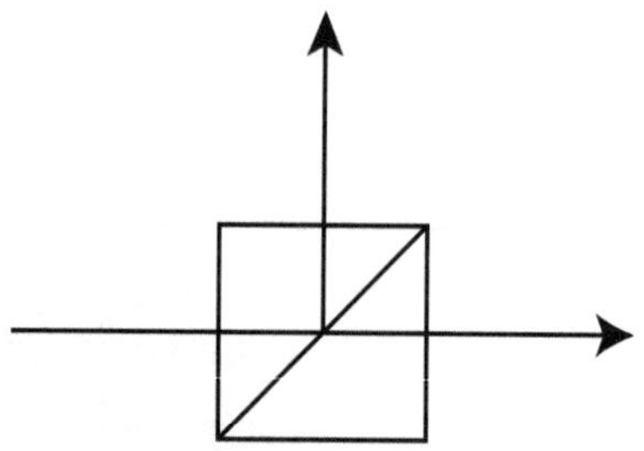

Figure 2.43

Beam splitter ray diagram.

coatings to operate in various wavelength and laser power regimes. You can also purchase beam splitters with different amounts of splitting – for instance, a 25/75 beam splitter might fork off 25 percent of your light intensity and transmit 75 percent of the original beam, whereas a 50/50 beam splitter cuts the power in two for both the transmitted and the split-off beams. Beam splitters come as cubes and as dielectric filter-like disks; both types can be polarizing or non-polarizing. Depending on the wavelength regime, size, and power rating, beam splitters can cost from $100 to $1000. We recommend purchasing beam splitters in a cube mount for easy handling and setup. If mounted in a kinematic cube mount, this allows for easy and fast removal/insertion of the beam splitter into the beam path for rapidly sending light to one component or another (see Figure 2.33). In a good beam splitter, the total intensity of the two outgoing rays is equal to the intensity of the incident beam. This may not always be the case; check with the vendor about an antireflection coating.

TIP: For a cheap ~5/95 beam splitter, you can use a glass slide positioned at 45 degrees in the beam path (see Section 5.8). We particularly recommend this trick for coupling white light into a beam path or splitting a small amount of laser light out of the beam path to send it to a power meter or a spectrometer to characterize your laser beam properties in real time during your experiment.

2.3.18 Filters and Dichroic Mirrors

Filters are essential to most optical measurements. Filters nowadays comprise stacks of thin films on a substrate of otherwise transparent material and are used to block or pass only certain wavelengths of light, known as **bands**.

Filters can be broadly divided into three classes: **neutral density**, **excitation**, and **emission filters**. Neutral density (ND) filters block light across a wide spectral range. They can be thought of as largely wavelength independent, whereas excitation and emission filters are used to only allow certain wavelengths of light to pass.

Neutral density filters can be reflective or absorptive. If you are using a reflective neutral density filter be very aware of where the reflected beam is going. The reflected beam can be quite strong and dangerous.

The terms **short-pass**, **long-pass**, **notch** (aka **blocking**), and **bandpass** (aka **clean-up**) describe the filter's transmission characteristics versus wavelength (see Figure 2.45). A short-pass filter will allow short wavelengths to pass through the filter and block long wavelengths. Vendors classify short- and long-pass filters by a wavelength, or multiple. These wavelengths are the points at which 50 percent of incident light is blocked by the filter (and 50 percent is transmitted). For instance, in the percentage transmission curves in Figure 2.45, filters would be listed on vendor websites by the wavelength at which % transmission equals 50 percent. How sharp the transmission to rejection transition is for a filter is referred to as the cutoff sharpness. The sharper the cutoff, the narrower the transition on a spectral plot from transmission to rejection. Some techniques such as Raman spectroscopy require very sharp cutoffs (≤ 1 nm spectrally). A notch filter will block only a specific range of wavelengths and will pass everything else. A bandpass filter will pass only a specific range of wavelengths and will block everything else.

Filters are often characterized by their **optical density** (OD) such that:

$$OD = -\log\left(\frac{\%T}{100}\right) \tag{2.1}$$

OD is described on a log scale so a filter with an OD of 1 will pass 10 percent of the light, a filter with an OD of 2 will pass 1 percent of the light, etc. A good filter can have an OD of 8, meaning that it will only pass 0.000001 percent of the light at the wavelength it was designed for.

Filters typically have labels on their edge describing their functionality. A filter labeled 515/20 is a bandpass filter with the center of the transmission at 515 nm and the full width of the transmission window is 20 nm: it will pass light from 505 to 525 nm. Since modern filters are made up of thin layers of dielectric material on a substrate, they are directional. Typically, there is an arrow on the edge of the filter indicating the direction. Usually the arrow is supposed to point in the direction that the light should propagate. Unfortunately, some manufacturers have adopted the opposite convention and the arrow points toward the light source. Always double-check which convention your filter manufacturer is using. Dielectric filters are also very angle dependent. Some filters are designed to work at normal incidence and some are designed to work at a 45 degree angle of incidence. Take care to keep track of what the filter was designed to do and be sure that you mount it at the proper angle, otherwise the wavelengths that are passed or rejected will be shifted.

TIP: When you are mounting a filter, mark on the outside of the mount which way the arrow on the edge of the filter is pointing. There is no way to determine which way the filter should face without taking the filter out of the mount again.

A short-pass or bandpass filter might be used to allow only excitation laser light, as opposed to white light or extra wavelengths emitted from your laser

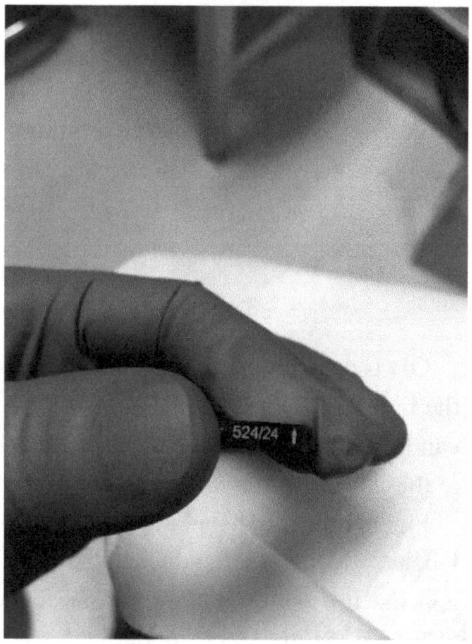

Figure 2.44

A bandpass filter viewed edge-on. The numbers depict the center of the transmission band and the bandwidth in nanometers. The arrow indicates the direction in which light should pass through the filter.

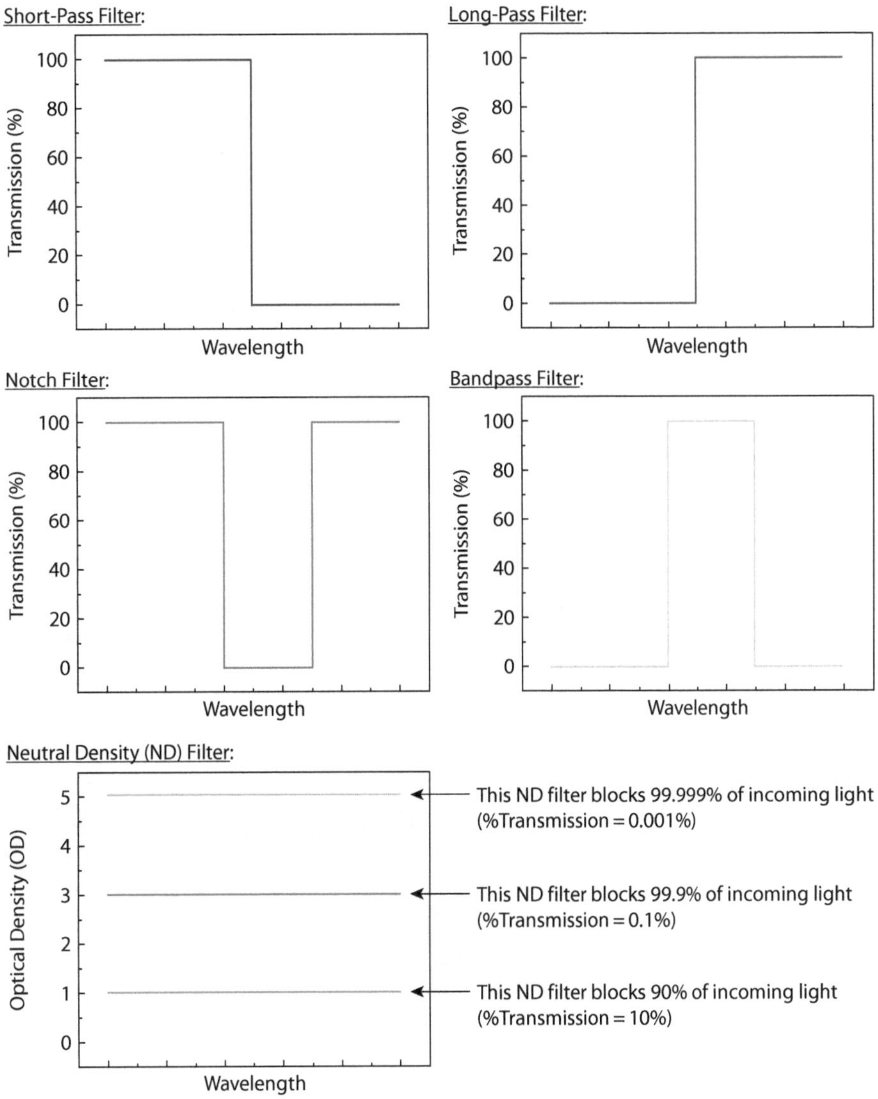

Figure 2.45

Typical transmission vs. wavelength characteristics of short-pass, long-pass, notch, bandpass, and neutral density (ND) filters.

(e.g., an undesired plasma line in a gas laser), to pass to a sample. Bandpass filters are also often used to allow only a specific emission wavelength to reach a detector or camera (e.g., for fluorescence microscopy, Section 4.7). Long-pass and notch-pass filters are most often used to block excitation (laser) light but allow emission from the sample to pass. All of these filters might be used to

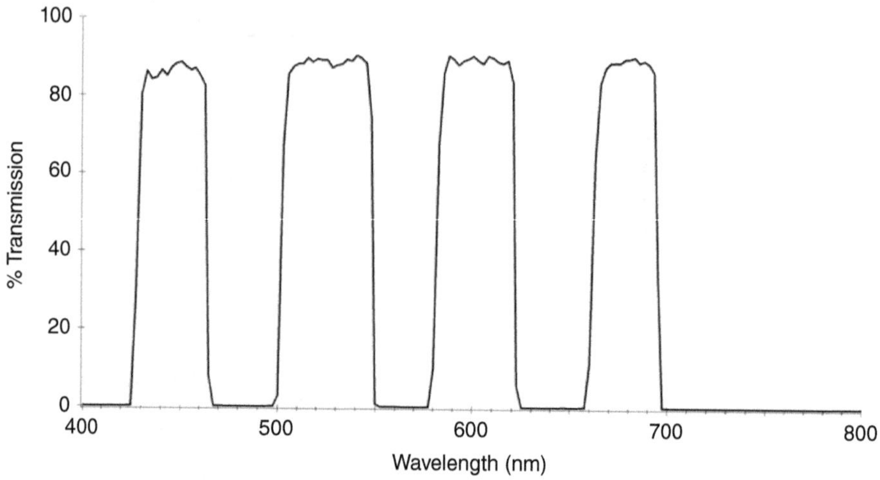

Figure 2.46

Transmission spectra of a custom quad bandpass filter for fluorescence microscopy applications.

Figure 2.47

Fixed 1" mount, filter, retaining ring, spanner wrench.

make a **broadband** light source useful as a specific color source by only allowing the wavelengths you want to pass through the filter. Note that just because a filter blocks light at a certain wavelength regime does *not* mean that that light is reflected – the light might undergo destructive interference as it interacts with the filter. Various notches or bandpasses can be combined in a single filter: quad-band filters are common in fluorescent microscopy setups.

Since filters are typically designed to be used such that the incident beam is normal to the filter surface, filters are usually mounted in a fixed mount.

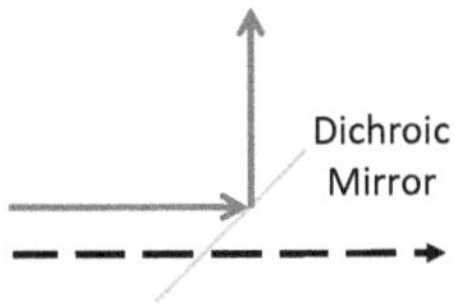

Figure 2.48

Ray diagram of a dichroic mirror mounted at 45 degrees to the incident light. In this case, this is a long-pass dichroic, so longer wavelength light (dotted line) passes straight through the mirror whereas shorter wavelength light (solid line) is reflected off of the mirror.

Dichroic mirrors are essentially the filter version of beam splitters – they are placed at 45 degrees in the beam path and reflect certain wavelengths but allow others to pass straight through. Dichroics are most often used to block laser light but allow emitted light from a sample of interest to pass to a camera or detector. Since dichroics need to be mounted at 45 degrees to the incident beam to work, we recommend mounting them as you would mirrors. Dichroics also often come in cube systems (see Figure 2.33) or custom mounts for use in microscopes.

Filters and dichroics can vary widely in cost from $100 to more than $2000, depending on diameter, mounting, power rating, and their optical performance; specific performance considerations include the steepness of the cutoff band, the percent blocking or transmission at your wavelength of interest (90 vs. 99.999 percent will be a big cost increase for some wavelengths), and the number of bandpasses or notches in the filter.

TIP: Optics labs seem to often end up with unlabeled filters lying around. Using a UV-VIS spectrometer allows you to determine the characteristics of an unlabeled filter.

2.3.19 Apertures, Irises, and Pinholes

An **aperture** is simply a hole through which light passes. **Irises** are adjustable apertures whose diameter can be adjusted; these are particularly useful for beam alignment and for cutting out imperfect **retroreflections** or **scattered light** (see Sections 5.2 and 5.3). Irises can be mounted simply by screwing the 8-32 set screw at their base into the appropriate end of a post.

Pinholes are literally just holes in small metal plates with diameters less than about 200 μm. Pinholes are used extensively in confocal microscopy (Section 4.9).

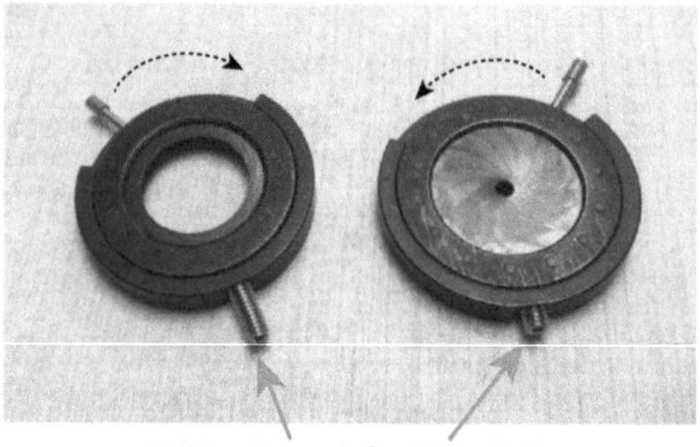

Figure 2.49

Adjustable irises. The iris on the left is open, the iris on the right is closed. The pin at the top of the iris slides to control the diameter of the opening.

Since pinholes are so small, we recommend using a magnetic snapping mechanism in a cage system to reliably center a pinhole in the beam path if you are building your own system.

2.3.20 Beam Blocks, Traps, and Shutters

Beam blocks and **traps** are designed to block or absorb incoming light. Simply place the block or trap in the beam path and it will absorb the incident light without allowing it to scatter. For CW light, a beam block is usually fine. Beam traps are slightly more expensive but are more versatile and can be used for CW and pulsed UV, visible, and IR light. For higher power lasers, we recommend a trap rather than a beam block – the block will scatter some light; this scattered light can still be harmful. Traps can get quite hot as they dissipate the energy in the laser beam, so use caution when touching.

Just like a camera's shutter, a laser **shutter** is simply a beam block that can flip open and closed. A beam shutter can be made by combining a piece of black paper, plastic, or metal (for low-power lasers) or a beam block/trap (for higher powers) with a magnetic or flip kinematic mount (Section 3.1.4), or simply a small servo motor attached to a piece of black cardboard. Fancier, computer-controlled shutters can be purchased for ~$500.

2.3.21 Lenses

Lenses are described by the shape of their cross-section. Lenses bend the path that light is traveling due to their shape.

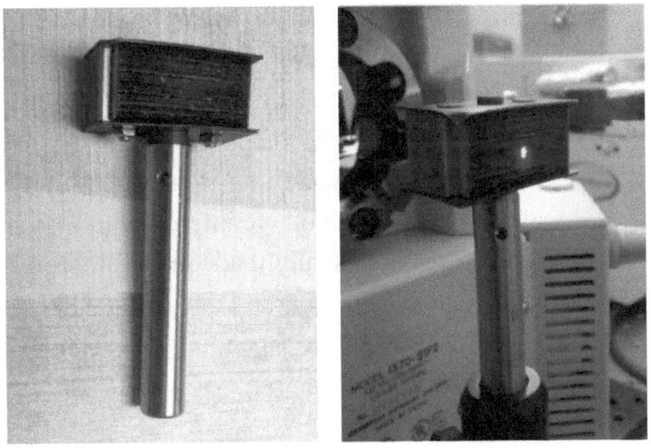

Figure 2.50

Beam blocks. An unmounted beam block is shown on the left. Right, a beam block blocking a 405 nm laser.

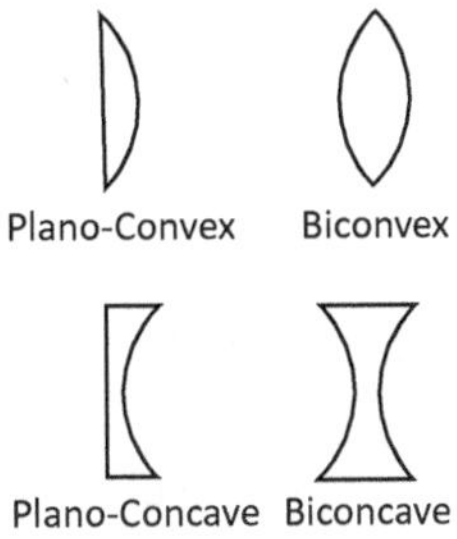

Figure 2.51

The cross-section shape for four common shapes of lenses.

The plano-convex lens is probably the most common lens in a lab. One side is flat (or plano) and one side is curved out (or convex). The next most common is probably a biconvex lens, in which both sides are curved out. A plano-concave lens has one flat side and one curved inward. A lens with both sides curved inward is a biconcave lens.

A plano-convex lens will take a collimated laser beam and focus it to a spot. A plano-concave lens will take a collimated laser beam and turn it into a diverging beam, which at first might not seem very helpful. The combination of a plano-concave lens and a plano-convex lens, however, will give you a beam expander without going through a tightly focused spot. See Section 5.1.9 for further details on building beam expanders.

Lenses are typically mounted in a fixed mount with a retaining ring. A special tool called a spanner wrench is needed to tighten and loosen the retaining ring (see

Figure 2.47). The spanner wrench has two little bumps that fit into the two little slots in the retaining ring. Please do not try to tighten or loosen retaining rings with any other tools. Trying to use a hex key (or tweezers or ballpoint pen) to tighten a retaining ring is pretty much a surefire way to scratch a lens.

Lenses can be made out of a wide variety of materials, depending on what wavelength of light you are interested in working with. For the visible wavelength range, lenses are usually made out of different types of glass. The UV range will need fused silica or quartz. For the mid-IR range, the most common materials are CaF_2, ZnSe, Si, or Ge, depending on exactly which spectral window in the mid-IR you are working in.

TIP: To estimate the focal length of a lens, stand under a ceiling-mounted fluorescent light fixture. Hold the lens above a piece of white paper and move it up and down until the image of the fluorescent bulbs comes into sharp focus (see Figure 2.52). The distance between the lens and the paper is the focal length of the lens. It can be estimated with a ruler at this point. Lens focal lengths are usually in units of millimeters (e.g., 25 mm, 50 mm, etc.).

2.3.22 Antireflection Coatings

Any time that light travels from a medium of one refractive index to another, there will be some reflection. The amount of this reflection can be minimized with an antireflection coating. You may be familiar with antireflection coatings as an option for eye glasses. The antireflection coating is made up of a series of alternating layers of high and low index of refraction materials. The exact thickness of these layers is important in determining over which wavelengths the antireflection coating is effective. The design of antireflection coatings is beyond the scope of this book. Antireflection coatings are readily available from lens manufacturers, so it is key to pick a lens with the appropriate coating for the wavelengths that you are using.

2.3.23 Lens Aberrations

Lenses are inherently imperfect. The imperfections in optical lenses are known as aberrations. We will discuss several of the most prominent aberrations here, but there are many additional types that are beyond the scope of this section. In glass

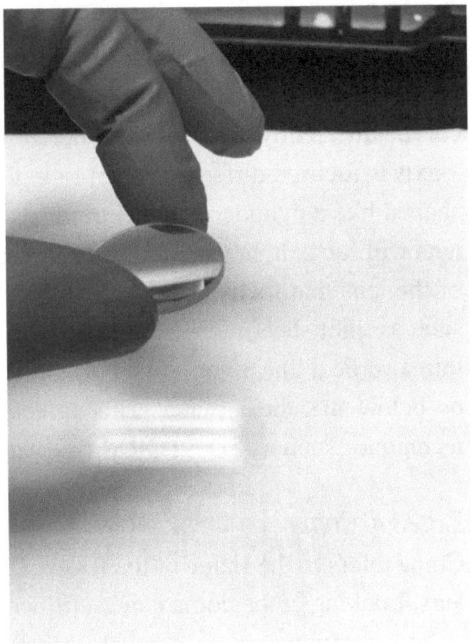

Figure 2.52

The focal length of an unmarked lens can quickly be determined by focusing the image of the ceiling light fixture. The distance between the lens and the image is the focal distance of the lens. Wearing gloves when handling unmounted optics prevents fingerprints on the optics.

lenses, there is not much you can do about aberrations, except to buy a better (more expensive) lens.

2.3.23.1 Chromatic Aberrations

Chromatic aberrations are due to the fact that the index of refraction of a material has a slight wavelength dependence. This leads to different colors of light focusing to different focal lengths, or distances beyond the lens. If you ever try to use a cheap toy microscope and you see different colors as you try to focus on the sample, you are seeing chromatic aberrations. The red light focuses to one plane and the green light focuses to a different plane.

2.3.23.2 Spherical Aberrations

Spherical aberrations refer to imperfections in the lens such that light passing through the edges of the lens focus to slightly different focal lengths than light passing through the center of the lens.

2.3.23.3 Astigmatism

Some readers may be familiar with the term astigmatism from visiting the optometrist for glasses. Astigmatism refers to the aberration where the lens is curved differently vertically and horizontally, so that light that enters a lens on its x-axis is focused differently than light that enters the lens along the y-axis. A lens shaped like a cylinder is the extreme example of an astigmatic lens. A cylindrical lens will focus light to a line instead of a point. Remember that it is the curvature of the lens that focuses the light. Cylindrical lenses are used in some techniques such as light sheet microscopy (Section 4.10). An astigmatic lens can focus light into a vertical line, a spot, or a horizontal line depending on whether you are above or below the focal plane. This effect has been used to good effect in some techniques such as 3D STORM (see Section 4.17) to encode z information.

2.3.23.4 Coma

Coma refers to the shape of the focused spot as the beam passes off-axis through a lens. Looking at the coma is a useful way of making sure that the beam is passing on-axis through a microscope objective. If the beam is not focusing to a circle but to a bright spot with a tail that looks something like a comet, you have coma. Observing coma usually means that light is not propagating through the center of the lens and you need to walk the laser beam (see Section 5.2) to realign your optical system.

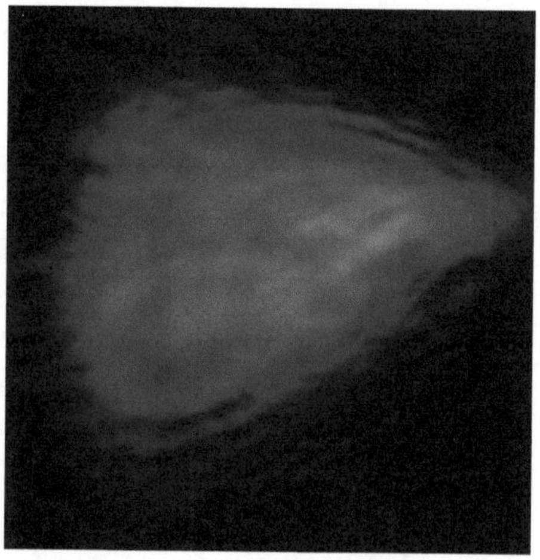

Figure 2.53

Image of focused laser spot showing coma due to the light not propagating down the center of a microscope objective, but instead coming slightly skewed down the microscope objective. Realigning the laser beam so that it is centered will help make the focused spot circular again.

2.3.24 Doublets

It is impossible to correct for all the aberrations in a single lens. The solution is to start to add more lenses to compensate for the different aberrations. The simplest version is a two-lens system often referred to as a doublet. Achromatic doublets – "achromats" – are two-lens systems that compensate for the chromatic aberrations seen in a single lens. These can be purchased from lens suppliers.

2.3.25 Microscope Objectives

A microscope objective is a package of numerous (~15) lenses chosen to minimize several different aberrations and give the best image quality. They do not necessarily need to be used in a traditional microscope body. Objectives have a threaded mount on their base for mounting. In a traditional microscope body the objectives are mounted in a rotating turret to allow for easy exchange of different magnification objectives. Threaded optical mounts for objectives are available so that the objective lenses can be mounted to posts and used in custom optical setups outside a traditional microscope body. There are several different standards for objective lenses' thread diameter and count per inch, so pay attention to make sure the objective and the mount will match. Thread adapters do exist to mate mismatched objectives and mounts.

There are quite a number of considerations to keep in mind when choosing a microscope objective. Since it is practically impossible to eliminate all aberrations

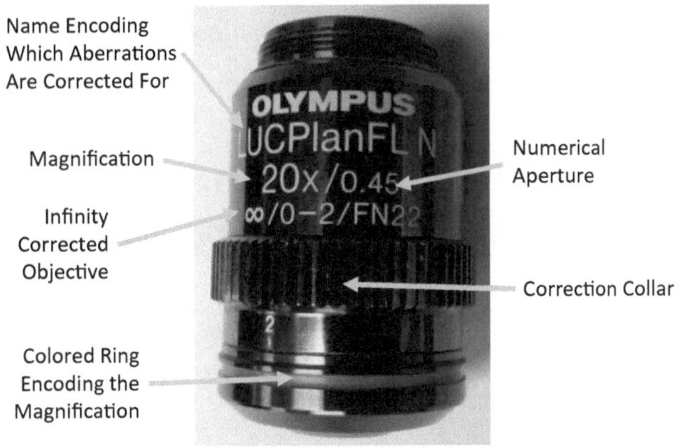

Figure 2.54

A long working distance 20× microscope objective.

at the same time, each microscope objective has some compromises built into its design. It is important to understand these compromises to get the best possible data to answer your scientific question.

2.3.26 Magnification

The value most people are familiar with for characterizing microscope objectives is the magnification. Typical values are 5×, 10×, 20×, 40×, 50×, 60×, and 100× magnifications. The magnification is always stamped on the side of the objective. It is also color coded in the colored ring around the objective (Table 2.2). This is a convenience so you can identify what the magnification of the lens is without needing to be able to read the label on the lens.

TIP: The highest magnification does not necessarily give you the best image. As you go to higher magnifications the field of view is smaller. You often collect fewer photons from a smaller field of view, which results in a dimmer image. Especially for fluorescent microscopy techniques, the sweet spot for image quality is often in the 40× to 60× range.

2.3.27 Field of View

The **field of view (FOV)** is how large an area can be imaged at the same time through an objective. The FOV is inversely related to the magnification. The larger the magnification of a lens, the smaller the FOV is.

Table 2.2 The color code conventions for microscope objective magnifications.

Magnification	Ring color
5×	Red
10×	Yellow
20×	Green
40×	Typically light blue
50×, 60×	Typically dark blue
100×	White

2.3.28 Depth of Focus

The **depth of focus** is the range in z, i.e., how thick a section of your sample is in focus at the same time. In general the higher the magnification the objective is, the shallower the depth of focus it has. This can be an advantage as it gets rid of portions of the sample above or below the plane you are interested in or it can be a disadvantage as many interesting samples have a lot of three-dimensionality to them and it would be great to be able to see everything in focus at the same time.

2.3.29 Infinity Corrected Objectives

Most microscope objectives that you will encounter today will have a ∞ on their side. This means that they are **infinity corrected** objectives. They take light from the focal spot and produce a collimated beam. Most modern microscope bodies are designed around this principle. Be careful of grabbing an old objective that is not infinity corrected and trying to use it in a modern microscope body.

2.3.30 Numerical Aperture

The **numerical aperture** (NA) is a measure of the solid angle over which the lens can collect light. The larger the NA, the better the lens is at collecting light.

$$NA = n \sin\theta \tag{2.2}$$

where NA is the numerical aperture, n is the index of refraction of the medium that is contacting the microscope objective, and θ is the angle that light is collected over.

Typical values for air objectives range from 0.05 to 0.95. You will observe dramatically increased costs for higher NA objectives. From Equation (2.2) the

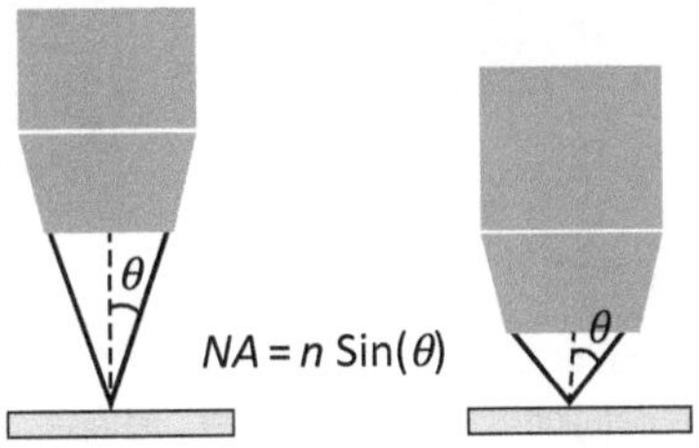

Figure 2.55

Numerical aperture of objective lenses.

physical limit when working with an air objective (where the index of refraction of air is $n \sim 1$) is an NA = 1.

We can change the medium contacting the lens to something with a higher refractive index such as water ($n = 1.33$) or oil ($n = 1.45$). This is done with specially designed objectives called immersion objectives (see Section 2.3.31).

2.3.31 Immersion Objectives

The numerical aperture of a lens can be greater than 1 if the index of refraction of the medium it is in contact with is larger than 1. Specially designed lenses called immersion objectives use media with indices of refraction greater than 1 to collect light more efficiently from the sample. Water ($n \sim 1.3$), which is useful for imaging biological samples, and oil ($n \sim 1.5$) are the two most common immersion media.

Despite the name, immersion objectives are *not* designed to be totally immersed in a liquid. Those types of objectives are referred to as dipping objectives (Section 2.3.32). Only a single drop of immersion oil or water is needed on the microscope objective. Use the oil sparingly as it can make a big mess quickly. The objective is then moved into contact with the cover glass of the sample and a small "bridge" is formed to help capture more of the light (see Figure 2.57).

(a)

(b)

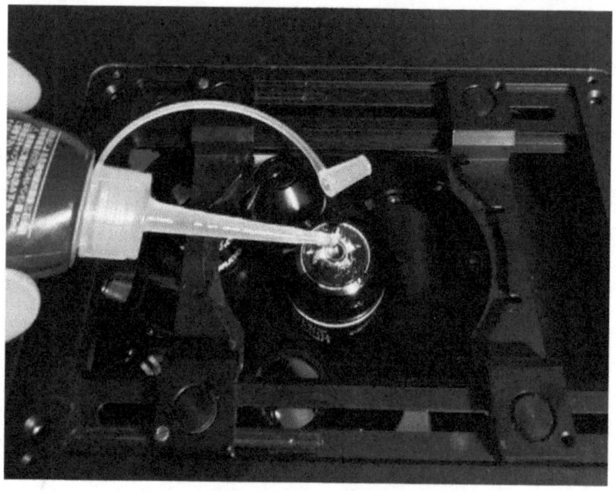

Figure 2.56

(a) Oil immersion objective with NA of 1.4. Note the lens is clearly labeled "oil." Putting oil on non-immersion objectives will just damage the objective. Only put immersion oil on objectives specifically designed for oil and clearly labeled oil. (b) Adding a drop of immersion oil to the oil immersion objective on an inverted microscope. Immersion objectives can be found on both upright and inverted microscopes.

Figure 2.57

Optical bridge formed by the drop of immersion oil between the circular microscope coverslip and the microscope objective lens.

Always, *always*, clean the oil off the objective when you are finished. The oil will dry if left overnight and the objective will be destroyed. A 100× oil immersion objective costs thousands of dollars, so this is a very expensive mistake to make. To clean the objective, gently wipe the excess oil off with a lens cleaning tissue. After the excess oil has been removed, place a couple drops of Sparkle™ glass cleaning solution or isopropyl alcohol on a second lens cleaning tissue and use this to wipe the objective clean of oil residue (see Section 5.1 for more about cleaning optics).

2.3.32 Dipping Objective

Some objectives for upright microscopes are designed to be totally dipped into the tissue culture media. These objectives are distinctive in having a ceramic or Teflon jacket around the lens (see Figure 2.58). This is different from an immersion lens, although the names are confusingly similar.

2.3.33 Working Distance

The working distance (WD) of an objective is the distance between the bottom of the lens and the top of the sample. It is typically not listed on the side of the objectives. For a standard objective it is often only a few millimeters. The higher the magnification of the objective, the shorter the WD will be. This can limit how deep into a sample you can image, or limit your experimental setup if your sample needs to be in a chamber such as a cryostat (see Section 2.3.57).

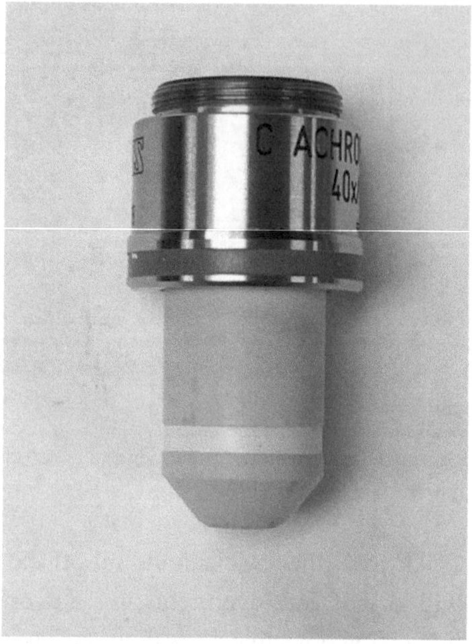

Figure 2.58

Dipping objective. Note the distinctive ceramic jacket to the objective. Other dipping objectives have a Teflon jacket. Dipping objectives are for upright microscopes and are designed to be dipped completely into tissue media, the liquid that contains nutrients for cultured cells.

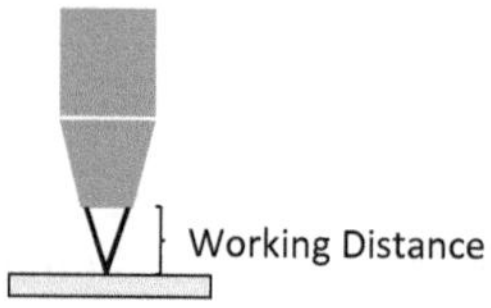

Figure 2.59

The working distance of the microscope objective is the distance between the bottom of the lens and the sample.

2.3.34 Long Working Distance Objectives

Sometimes it is desirable to be able to keep the lens further away from the sample, such as when a sample is in a cryostat. Special lenses known as **long working distance (LWD) objectives** have working distances of 0.5–1 cm (e.g., 15 mm for 20×, 12 mm for 50×, 6 mm for 100×). The NA for such lenses is much smaller. For example, a typical 100× air objective would have an NA of 0.9, whereas a LWD 100× would have an NA of 0.6.

2.3.35 Correction Collar

There is sometimes a rotating ring on an objective (see Figure 2.54). This could be a correction collar. It adjusts the distances between some of the internal lenses in the objective to allow you to compensate for different thicknesses of glass (e.g., glass cover slips) covering your sample. To get a good image it is essential to properly adjust the correction collar. Note: Not all external rotating rings on microscope objectives are correction collars. Sometimes the rings control an aperture.

2.3.36 Diffraction Gratings

A **diffraction grating** is a series of fine parallel lines scratched into a piece of glass or plastic. When white light hits the series of lines it will undergo wavelength dependent dispersion and spread out like a rainbow. You are probably familiar with diffraction gratings as the rainbow glasses in the gift stores of science museums.

Diffraction gratings allow you to sort the photons in your beam by color and are the basis of most dispersive spectrometers. They can be designed to work in transmission or reflection. The reflection type gratings are more prevalent in scientific instruments, though transmission spectrometers are used in certain applications.

The **blaze** of the grating is the number of lines per millimeter. Sometimes they are also referred to as grooves per millimeter (gr/mm). The denser the line spacing, the bigger the angle of dispersion (the more the rainbow is spread out). For historic reasons standard line densities are 150, 300, 600, 1200, 1800, and 2400 gr/mm.

Diffraction gratings are designed for a specific wavelength. They are said to be "blazed at 532 nm" for instance. Try to pick a grating that is blazed for near the center of the wavelength range that you are using. Diffraction gratings often have surprisingly low throughput efficiency (e.g., 30 percent throughput efficiency is often considered pretty good for a diffraction grating) even at the wavelengths that they are designed for.

TIP: *Never, ever* clean a diffraction grating with anything other than compressed air. The lines on the grating are very soft and will be destroyed by contact with lens paper.

Note that, in reality, diffraction gratings spread incident light into a series of successively dimmer and dimmer rainbows. This series of rainbows is known as

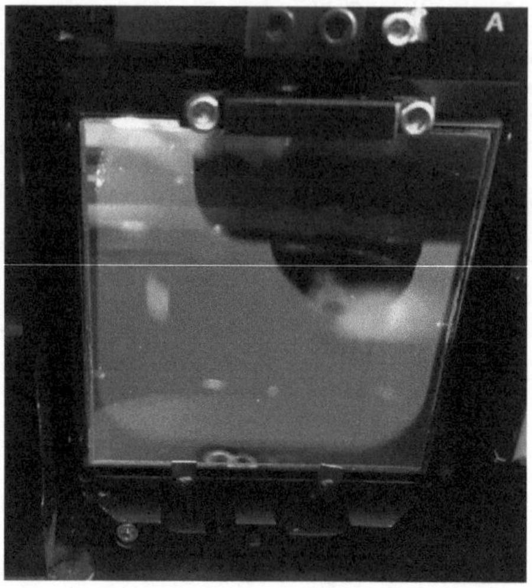

Figure 2.60

Diffraction grating mounted in a spectrometer.

the orders of the diffraction grating. It is this series of rainbows that makes the throughput on diffraction gratings so low, as many of the incident photons contribute to the higher orders of the grating. The zero order of a reflective diffraction grating means that the grating is at an angle such that it acts as a mirror and doesn't spread out the light. The first order is the brightest rainbow, so it is typically used for spectroscopy applications. If you scan too far you can hit the second order diffraction peak and see a large peak from the excitation light. This can be surprising and misleading if you are not paying attention. If you see a large peak at around twice the excitation light wavelength you are probably seeing the second order diffraction of the excitation light. If you need to scan in that spectral range you will have to add an optical filter to cut the excitation light before sending the signal to the spectrometer.

2.3.37 Fiber Optics

An **optical fiber** (aka **waveguide**) is a glass fiber through which light travels. Typically, these fibers are flexible coaxial (core/shell) cables where the core material has a lower refractive index than the shell. Thus, light is transmitted down the cable via **total internal reflection** (Section 1.8), so they provide a reliable and fairly efficient way to get light from one spot to another. Since fibers are flexible and can be purchased in various lengths, the start and end of your

Figure 2.61

A microscope objective lens being used to couple light into a fiber optic.

cables can be in creative and strange places such as neighboring rooms, into a microscope, or in a different *z*-plane on your laser table. In fact, fiber optic cables are used in undersea cables to transmit cat videos from North America to Europe, in endoscopes which doctors use to check out your guts, and in mining applications to couple high-power lasers into deep bore holes for blasting away rock and mud.

Coupling refers to linking an optical fiber to a component such as a laser; one might have a fiber-coupled laser which, instead of emitting a beam into free space, emits its light into a fiber. Coupling is often "lossy" – that is, some light intensity will be lost via scattering when the light enters and leaves the fiber. The power loss from source to fiber is quantified by **coupling efficiency**.

The disadvantage of optical fibers is that it's harder to put components like filters, polarizers, and waveplates into the beam path, especially since power is lost when light enters and exits the fiber. However, one can purchase collimators, couplers, and isolators for fibers which help to minimize these losses. **Collimators** keep exiting light beams well collimated; **couplers** allow more light to enter the fiber and can also be used to link two fibers together; and **isolators** only allow light to travel one way into a fiber (thus preventing interference from, say, reflected light traveling the wrong way down a fiber cable). **Single-mode fibers (SMFs)** only efficiently transmit light down the fiber via the "transverse mode." **Multi-mode fibers (MMFs)** are thicker than SMFs so that they can support multiple propagation modes. Fibers can also be purchased that maintain the polarization of the incoming light. Components such as couplers, collimators, and isolators can be quite expensive, often up to several thousand dollars per component for, say, polarization-maintaining, multi-mode couplers.

Fibers can also be coupled to **optical amplifiers** – this might consist of a fiber containing rare earth ions. Analogously to laser gain media, the rare earth ions (Er^{3+}, etc.) inside the fiber core might fluoresce when they're hit with light in the fiber, thus emitting more light to compensate for loss in the cable. Likewise, a **fiber laser** simply uses an optical fiber doped with rare earth ions as a gain medium.

2.3.38 Spectrometers (Monochromators)

A spectrometer and a monochromator are both used to sort photons by color and are very similar in design. The primary difference is what is done with the photons after they are sorted. In a spectrometer, the sorted photons are detected and a spectrum is recorded. In a monochromator, a single desired color is sent through the output port and used in the next stage of the experiment.

There are a number of spectrometer designs, but the most common design for spectrometers and monochromators is known as the Czerny–Turner configuration (see Figure 2.62). First, light from an **entrance slit** is focused by a curved mirror onto a diffraction grating. The diffraction grating then spatially separates ("**disperses**") different colors of light. The dispersed light then bounces off another curved mirror which forms an image that is smeared out horizontally by the wavelength of the slit on the detector. On the detector, the vertical direction of the image still represents real space dimensions (i.e., up–down), but the horizontal axis represents wavelength. The vertical direction is usually **binned** (i.e., all the pixels in the same column are added together) to generate a stronger

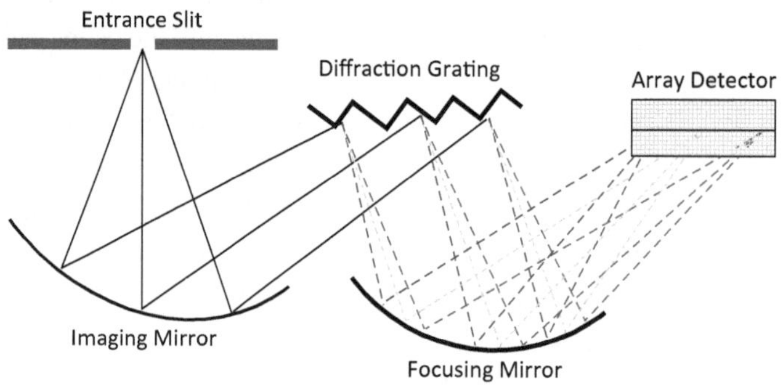

Figure 2.62

Czerny–Turner spectrometer layout.

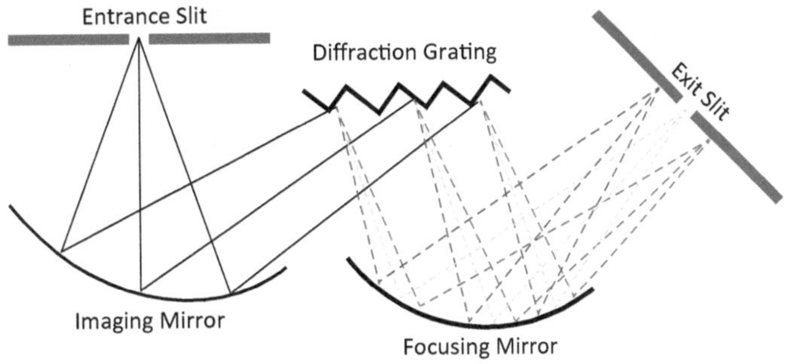

Figure 2.63

Czerny–Turner monochromator layout. Polychromatic light enters and monochromatic light exits.

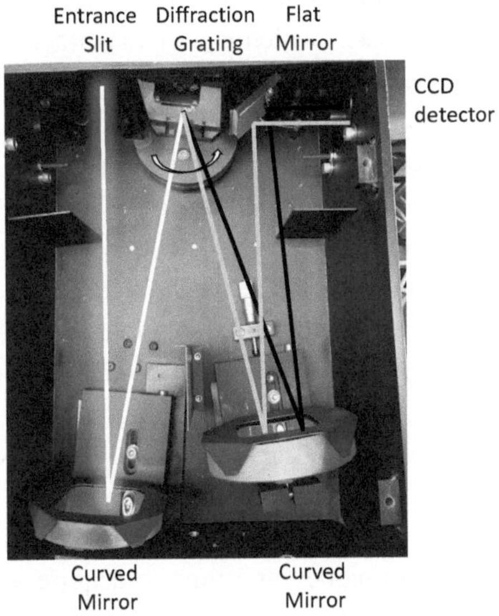

Figure 2.64

View of the inner workings of a spectrometer. Light enters the entrance slit. The entrance slit is then imaged onto the diffraction grating with a curved mirror. The diffraction grating spreads out the light by wavelength like a rainbow. The rainbow is imaged onto the CCD detector by another curved mirror. Pixels on the right of the camera will count the number of blue photons, and pixels on the left of the camera will count the number of red photons.

signal, but the vertical direction doesn't have to be binned in the case of line scanning spectral imaging systems. Czerny–Turner spectrometers typically are not very good for imaging applications. For spectral imaging applications, a spectrometer that is designed to compensate for the optical aberrations of the mirrors and diffraction grating is important.

As the name suggests, monochromators transmit only a certain color of light. A typical ray diagram for a monochromator is shown in Figure 2.63. At the entrance to a monochromator, you will find a narrow entrance slit. Light passes through the slit and reflects sequentially off of a curved mirror, a diffraction grating, and another curved mirror. The **exit slit** is then used to select which wavelength of light will be passed on to the next stage of the experiment. The diffraction grating typically is rotated to select different wavelengths of light. A cheaper way to make a spectrometer without an array detector combines a monochromator with a narrow exit slit with a photodetector. Computer coordination between the position of the diffraction grating and the intensity of light recorded by the photodetector can yield a spectrum. This slows down your data acquisition rate significantly though, so most spectrometers these days use an array detector or CCD detector.

There are a number of factors that determine the spectral resolution of a spectrometer, which is why manufacturers are so reluctant to quote a number for the spectral resolution. Some of these factors include the width of the entrance slit, the blaze of the grating, the distance the light is allowed to travel after the diffraction grating, and the pixel size of the detector.

The smaller the entrance slit width, the higher the spectral resolution, but the lower the signal level. The higher the blaze of the grating, the higher the spectral resolution, but the lower the signal level. The longer the spectrometer path length, the higher the spectral resolution, but the more sensitive the spectrometer is to small temperature fluctuations in the lab. The smaller the pixel size, the higher the spectral resolution, but the lower the signal level. In steady-state experiments you can often compensate for the lower signal levels by acquiring photons for a longer time (aka longer acquisition or "**exposure times**"). If, however, you don't need the extra spectral resolution to answer your scientific question, we recommend not spending the extra money and time to acquire a higher-resolution spectrum. This is why spectrometers often come with two or three user exchangeable diffraction gratings to allow the user to balance their need for spectral resolution with acquisition time.

TIP: The throughput and efficiency of spectrometer components, particularly diffraction gratings and detectors, are not equal over all wavelength ranges. If you need to quantitatively compare the intensity of one wavelength signal to another, be sure to collect a baseline instrument response from a reference sample and normalize your experimental data by the wavelength-dependent throughput and efficiency of the various components as measured from your reference sample. Similarly, monochromators have wavelength-dependent transmission efficiencies, so the power of the input light at various wavelengths may not necessarily be equal or even uniformly proportional to the output power at the same wavelengths. Double-check the wavelength-dependent properties of your components before you begin collecting data, so you know how to normalize your spectra later on.

2.3.39 Detectors

The underlying principle of photodetectors is to use a photon to excite an electron across a bandgap in a semiconductor material and then count the number of excited electrons to determine how many photons there were.

No detector is perfect at converting all the photons into excited electrons. How good a detector is at turning photons into electrons is known as the **quantum efficiency (QE)** of the detector. There is a wavelength dependence to the quantum efficiency for a detector. Some detectors have QEs greater that 95 percent at certain wavelengths. Just keep in mind that suppliers' specification sheets describe what the maximum QE for a detector is, but that may not be relevant if you need to detect light at a different wavelength. Manufacturers have QE curves showing the QE at different wavelengths for their detectors; you may have to request this however.

Electrons can also be excited thermally. These thermally excited electrons are known as the **dark counts** – the number of counts the detector reports even when there is no light on the detector. To minimize the dark counts, many detectors are cooled. Some are cooled with a thermoelectric (TE) cooler, which can cool a detector reliably to –70 °C, or with liquid nitrogen which can cool a detector to –196 °C (77 K). The rule of thumb is that for every five degrees Celsius that you cool the detector, you cut the dark counts by half.

One rule of caution about cooled detectors is that you do need to be careful about water condensing on chilled surfaces. This is a particular problem in hot, humid environments, since water condensing on cold surfaces or cooling lines can short-out electronics.

2.3.40 Point Detectors

A detector that has a single light-sensitive element is known as a point detector. Point detectors provide no spatial information about where the light came from. They simply count how many excited electrons were generated over a certain time period.

Figure 2.65

Silicon photodiode. The small square at the center is the actual photodiode.

2.3.41 Photodiodes

Photodiodes are single-element semiconductor detectors that absorb photons by exciting an electron across the bandgap of the material. Typically a voltage is measured as the output signal, which will be proportional to the incident light. Different semiconductors will be sensitive to different wavelengths of light. Silicon is typically sensitive between 200 and 900 nm. Indium gallium arsenide (InGaAs) is typically sensitive from 900 to 1700 nm.

Photodiodes can be used to make time-dependent measurements. The rule of thumb is: the smaller the diode the faster the response can be. A typical photodiode can be as fast as 10 ns. Avalanche photodiodes (APS) are very fast photodiodes, but they are only good for short time measurements (500 ps to 2 ns for example) and can be damaged by continuous illumination.

2.3.42 Photomultiplier Tube

Photomultiplier tubes (PMTs) were the gold standard photon detector for many years and they are still found in many systems. They have the advantage of having a variable amplification factor to increase their sensitivity when needed. They are also comparatively cheap (~$3,000) as far as detectors with single-photon sensitivity go. Photomultiplier tubes consist of a bialkali (e.g., CsRb alloys) coated on a metal electrode in a vacuum tube. Several additional metal electrodes, known as a dynode chain, sit behind the first CsRb-coated plate. Several hundred volts are typically applied across the dynode chain. When a

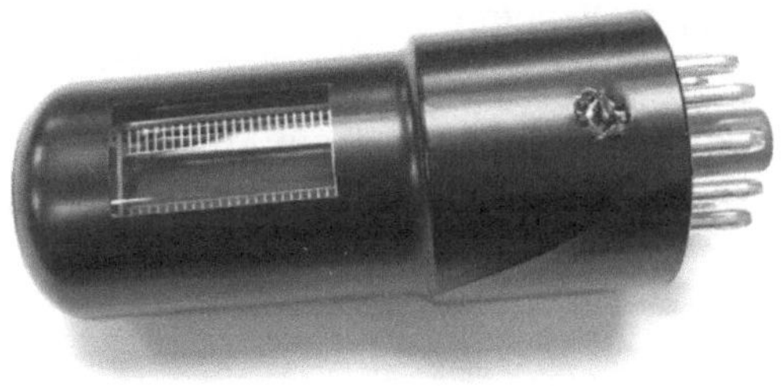

Figure 2.66
Photomultiplier tube (PMT).

photon hits the bialkali metal and ejects an electron due to the photoelectric effect (Appendix 1), the applied voltage accelerates the electron down the dynode chain, making more electrons (and thus amplifying the signal) at each dynode. PMTs typically have three settings that you can adjust: the voltage, the gain, and the offset. You typically want to just work with the voltage. Increasing the voltage will increase the amplification factor for each photon. Increasing the gain setting will increase the signal but it will also increase your noise. Changing the offset changes where the zero point is on your photon counting scale. If you plan to do any semi-quantitative measurements, you need to leave your gain and offset alone and always use the same PMT voltage.

2.3.43 GaAsP Detectors

Gallium arsenide phosphide (GaAsP, often pronounced "gasp") detectors are starting to replace PMTs in the visible light region. They are more sensitive and less noisy for about the same price as a PMT. The drawback to the GaAsP detector is that it has no sensitivity in the far red to near IR spectra region. Be aware of which spectral region you are planning to work in.

2.3.44 Mercury Cadmium Telluride Detectors

Mercury cadmium telluride (MCT) detectors are standard detectors for the IR region, especially beyond about 2 μm in wavelength. Mercury cadmium telluride is a ternary semiconductor compound that has a tunable bandgap depending on the exact composition. These detectors usually need to be cooled. They can be cooled either with a thermoelectric (TE) cooler, also known as a Peltier cooler, or with liquid nitrogen. They are frequently used in FTIR spectroscopy.

2.3.45 Array Detectors

Array detectors consist of a number of point detectors positioned next to each other. Array detectors are sometimes used in spectroscopy where the information of interest (e.g., wavelength) is spread out spatially like a rainbow in one axis.

2.3.46 Photodiode Arrays

A photodiode array is simply a line of photodiodes next to each other in order to gain some spatial information. Simple spectrometers often have a photodiode array as a detector.

2.3.47 Charge Coupled Device

A **charge coupled device (CCD)** is the technical term for a common class of digital cameras. It consists of a two-dimensional array of detectors called pixels. A good research-grade microscope CCD camera costs about $15,000.

CCDs can be optimized for different wavelengths in the visible and UV spectra (200–900 nm is the typical operating range). Since CCDs are typically comprised of Si, the quantum efficiency of CCDs usually drops off after 900 nm and no light can be detected beyond 1100 nm. CCDs are typically used for spectroscopy and imaging, and are often cooled to –70 °C with TE coolers.

> TIP: Scientific grade CCDs are sensitive to even low levels of light, so it is of utmost importance that you do not aim a high-power laser beam into a CCD. We suggest placing a notch filter centered on your laser wavelength in front of the CCD. If you need to image the laser spot for alignment, we recommend installing a second, less sensitive and cheaper camera at 90 degrees to the beam path. You can easily insert or flip up a mirror on a kinematic flip or magnetic mount aligned at 45 degrees to the beam path to reflect the laser beam into the cheap camera. This way, you avoid any risk whatsoever of your laser beam frying your expensive and delicate CCD.

Most CCDs have an option to bin during data acquisition. **Binning** consists of summing multiple pixels in a group (the pixels are "binned") on a camera together. In essence, binning combines small pixels into larger pixels to yield a larger signal. Naturally, though, this binning reduces either spatial or spectral resolution

Figure 2.67

Camera chips come in different shapes. Long narrow formats are used for spectroscopic applications. Square formats are used for imaging applications.

in the axis in which the pixels are combined. Therefore, in spectroscopy, it is common practice to use a CCD as the detector, but to bin the pixels vertically to create a virtual array detector.

2.3.48 Electron Multiplying Charge Coupled Device

An **electron multiplying charge coupled device (EMCCD)** basically has an electronic amplifier attached to each pixel. This makes the camera very sensitive and it can be used in applications where you need to count single photons. An EMCCD camera will usually cost upwards of $30,000.

EMCCDs can have quantum efficiencies as high as 95 percent at certain wavelengths. As EMCCDs follow a similar detection scheme to a normal CCD, they too can be optimized for different wavelengths in the visible and UV spectra (200–900 nm is the typical operating range); being silicon-based, their quantum efficiency typically starts dropping after 900 nm and no light can be detected beyond 1100 nm. Electron multiplying charge coupled devices are even more sensitive than typical CCDs, so heed the above warning closely to prevent burning out your expensive detector!

2.3.49 Scientific Complementary Metal Oxide Semiconductor

Scientific complementary metal oxide semiconductor cameras use a slightly different electronics architecture than CCDs, but the details are beyond the scope of this book. They are a more recent camera architecture for the scientific community, starting ~2010. Interestingly, SCMOS cameras have the same electronic design as the camera in your cell phone, and cell phone camera development helped push the technology to the level where it can be used for research-grade cameras.

These cameras are typically more expensive than CCDs but cheaper than EMCCDs and currently about $20,000. They typically offer a larger FOV and are a little quieter than CCDs. At the moment, many microscope designs can't actually take advantage of the larger FOV and the edges of the camera chip are usually not used in image acquisition. This is rapidly changing though as manufacturers redesign their microscope optical paths to take advantage of the capabilities of the new detector. So just be aware that adding a SCMOS camera to an old microscope body will not get you as much benefit as adding a SCMOS camera to a newly redesigned microscope body.

2.3.50 Streak Cameras

Streak cameras are very fast, very expensive (~$100,000) cameras that offer extremely fine time resolution (nanoseconds to picoseconds). In streak cameras, a photon hits a micro channel plate and an electron is ejected. The electron flies between two charged plates before hitting a phosphorescent screen. The voltage on the plates is swept so that electrons that are launched at different points in time will land on different parts of the phosphorus screen. When the electrons hit the phosphorus screen they cause photons to be emitted. The photons from the phosphorus screen then are imaged onto a normal CCD (or EMCCD).

Streak cameras have the advantage of capturing temporal information about the emission of light from a single photoemission event from your sample. If your sample is precious and will be obliterated in the single event, then the streak camera is the way to go. If your sample can be repeatedly excited and the photoemission events are all equivalent, there are cheaper ways to collect time-resolved information about your sample, such as using time correlated single-photon counting electronics for fluorescence lifetime imaging (Section 4.12) or using pump–probe spectroscopy techniques (Section 3.8).

2.3.51 Alignment Tools: Cards, Viewers, and Targets

On a typical optics table, you will find a variety of business cards scattered around. These cards are for sticking into the laser beam to check alignment. Since properly specified laser goggles prevent you from seeing the laser beam, you'll need a tool to align it even when you are wearing goggles. So, you can use a card that fluoresces when it is excited by the beam, and emits light at a wavelength that is not blocked by your laser goggles. Every laser jock has a favorite business card. The different papers in many business cards typically have fluorescent dyes (known as optical brighteners) that will be compatible with different laser/goggle combinations. Fluorescent colored index cards and Post-it/sticky notes from an office supply store often work well too. Optics companies also sell laser viewing cards for seeing the laser beams through laser safety goggles for a variety of wavelength regimes, ranging from IR to UV. We recommend trying out several cards to see what will work. One safety note: if you are using a high powered laser (e.g., pulsed or UV), be sure that the card can handle the laser power. If not, you might singe or ignite your alignment card. We've made that mistake several times with our pulsed systems. The soot can be bad for optics.

2.3.52 Acousto-Optical Tunable Filter

An acousto-optical tunable filter (AOTF) is a crystal that can be tuned to transmit a certain wavelength of light based on the radio frequency (RF) pulse that is sent into the crystal. The RF pulse changes the spacing of the crystal lattice and the other undesired wavelengths of light are diffracted out of the optical path. They are typically used to rapidly switch between different wavelengths from a multi-line laser or a broadband light source.

2.3.53 Acousto-Optical Modulators

Acousto-optical modulators (AOMs) are similar to AOTFs, except that they are used as a fast switch to modulate the intensity of the laser that passes through the crystal. The light of a single wavelength is diffracted in and out of the optical path, resulting in a modulated laser intensity and a known frequency. They are often used in conjunction with a lock-in amplifier. The lock-in amplifier only looks for a signal at the modulation frequency of the excitation laser, allowing weak signal to be extracted from the noise.

2.3.54 Electro Optic Modulators (EOMs)

Electro optic modulators (EOMs) are similar to AOMs except that they use high-voltage electrical pulses instead of acoustic pulses to modulate the diffraction of the light through the crystal.

Figure 2.68

Electro optic modulator (EOM) in an optical setup.

2.3.55 Diffusers

Diffusers are usually flat ground glass that spread out incident light in a fairly random manner. They are often used in microscopes to remove patterns from light sources such as the image of the filament from a tungsten lamp or speckle in a laser beam.

2.3.56 Nonlinear Crystals

Nonlinear crystals are a special class of materials that absorb two photons and emit one, or absorb one photon and emit two. In both cases, energy is conserved, so the energy of the input photons has to equal the energy of the output photons. For instance, you can send two photons in and get one photon out that has the sum of the energy of the two input photons. Or you can send in one higher energy photon and split the energy of the original photon between the two output photons. The split does not have to be even; in some cases, you can get two different wavelengths out. This is roughly the basis for optical parametric oscillators (Section 2.2.7).

Processes that involve multiple photons such as the above are known as **nonlinear optical processes** (see also Section 4.11). The probability of these nonlinear processes is low, so they typically only work with focused, short-pulse lasers (typically ~10 ps or shorter) where you have enough photons at the same point in time and space that the probability of two photons interacting simultaneously with the crystal becomes finite. Barium borate (BBO; "bee-bee-oh") crystals are a common frequency-doubling crystal found in labs.

2.3.57 Cryostats

In solid state physics, temperature is the easiest thermodynamic variable to control. (Pressure can be controlled with diamond anvil cells.) **Cryostats** allow you to cool samples and measure how the physical properties of the sample change. Cryostats often have optical windows so that low-temperature spectroscopy experiments can be performed. It is important to take a moment to think about whether the window material is transparent in the wavelength ranges that you are interested in. Cryostats can operate at liquid nitrogen (77 K) or liquid helium (4 K) temperatures. These ultra-cold liquids are known as **cryogens**. Your sample can sit either in the vapor of the cryogen or under vacuum. In the latter case, the sample needs to be mounted on a heat exchanger that is in contact with the cryogen outside of the vacuum to conduct heat from the sample to the cryogen. Often, a resistive heater is embedded in the sample mount and used to counter the cooling of the cryogen to allow measurements to be made at above that of the cryogen itself. Temperature controllers contain electronics that can regulate the load applied to the resistive heater to balance the cooling, providing tight control of temperature or temperature ramps. Older-style continuous flow cryostats, in which the cryogen is released to the lab after it passes by the sample, are going out of fashion as the world faces helium shortages. New, better systems are closed cycle, which means that they capture the helium and reuse it locally.

TIP: Unwaxed dental floss remains stable at low temperatures and has low outgassing, so using it to secure samples, stray wires, etc. inside the cryostat is sometimes a useful trick.

Safety note: cryogens are extremely cold and can give you frostbite. Be sure to use proper personal protective equipment (e.g., a face shield and a cryo-compatible apron and gloves). Also, as cryogens boil, they expand in volume by a factor of ~700. Therefore, a leak can quickly displace all of the oxygen in a small space, so be sure to have good ventilation and air flow to avoid suffocation.

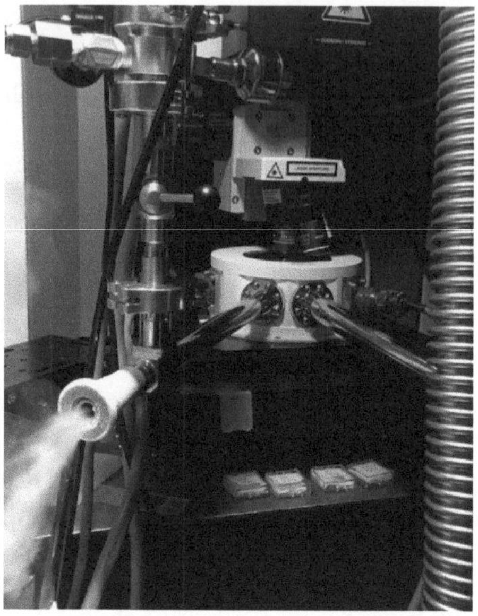

Figure 2.69

A cryostat in action underneath a microscope. Excess nitrogen is being vented to the room. Vacuum lines, temperature control cables, and cryogen lines are all connected to the cryostat.

2.4 Cell Phone Optics

Cell phones, and their cameras, have become ubiquitous in the last decade and are consequently making their way into the scientific lab. Therefore, it is worth dedicating a section to optical components that one can buy for cell phones. Although these cell phone-compatible optical components are basic, they are often extraordinarily affordable, simple, and robust, and the functions available are growing every day, often through do-it-yourself (DIY) projects.

2.4.1 Microlenses and Spectrometers

Cell phone cameras are meant to image macroscale objects such landscapes, cats, and your own face for selfies, but simply adding a lens can often turn your cell phone camera into a reasonable-quality microscope, or even spectrometer. A variety of options are available, but since the internet is a dynamic and evolving place we won't try to provide links to a specific option. Our point is that the camera on the cell phone in your pocket is a pretty good photodetector and shouldn't be overlooked. A variety of DIY websites exist for ideas such as www.instructables.com and www.publiclab.org.

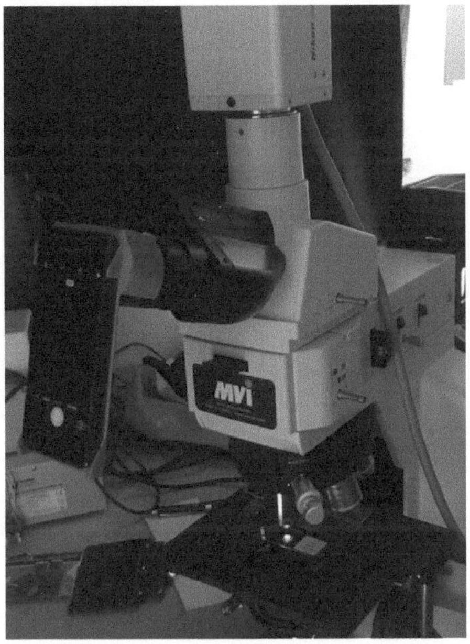

Figure 2.70

A 3D-printed eyepiece adapter to mount a cell phone camera onto a microscope for image capture.

2.4.2 Eyepiece Adapters

Cell phone camera quality has dramatically improved, so they can often be used to record photos from larger microscope assemblies instead of using a more expensive CMOS detector. For instance, a 3D-printable eyepiece adapter to mount your cell phone on the eyepiece of a microscope can be found on Thingiverse.com.[1] Commercial solutions are also available.

2.4.3 Infrared Imaging

Typically, it is desirable to block IR light for photography applications to mimic the responsivity of the human eye, so cell phones have short-pass filters in their cameras to block IR. But, if you remove this filter, cell phone cameras, which are comprised of Si chips that are sensitive to near IR light, can image in the near IR to give you some semblance of night vision. (Most webcams and digital cameras

[1] www.thingiverse.com/thing:59344

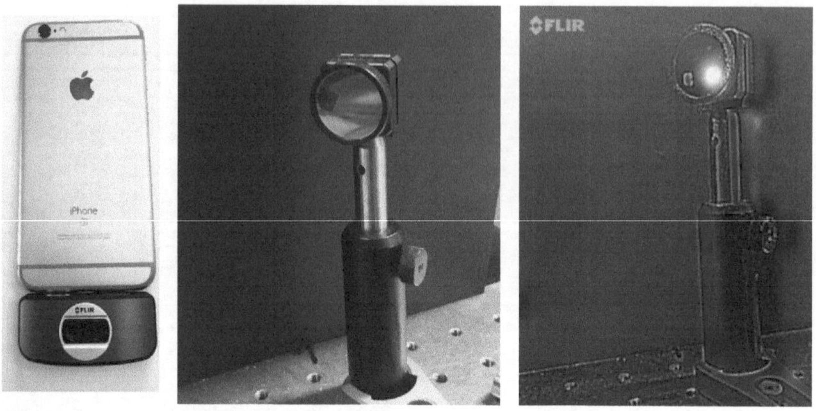

Figure 2.71

FLIR infrared camera attached to an iPhone; visible light image of mirror with infrared laser hitting it; infrared image showing where the laser is hitting the mirror.

have infrared filters, too, so the same "hack" will work for those devices.) A variety of instructions are available on the internet.[2]

If combined with a small spectrometer, you can also turn your cell phone into a cheap IR spectrometer that can be used to monitor plant health or determine the chemical makeup of food. Commercial cell phone adapters that do not require taking apart your camera for IR imaging are available, too.[3] The Raspberry Pi NoIR camera already has the IR filter removed.

2.5 Further Reading on Optical Components

Many optics vendors have useful application notes and specification sheets about the optical components they sell on their websites. The application notes and specification sheets tend to be written assuming a fair amount of prior knowledge. Hopefully after reading this chapter you have enough prior knowledge to understand the application notes from the vendors.

2.6 Vendors

This section is by no means exhaustive or comprehensive. It is a reflection of the experiences that the authors have had in working in various optics labs for the past 15 years.

[2] www.instructables.com/id/Cell-Phone-Night-Vision-Under-10/?ALLSTEPS

[3] www.flir.com/flirone/ios-android/

General: These vendors sell everything you need for an optics lab: Edmund Optics, Newport, Thorlabs. Thorlabs is probably the most popular as they have a comprehensive catalog that is easy to navigate and they ship very quickly. On the east coast of the United States you can usually have the part you need in the lab 36 hours after you order it.

Filters: Semrock and Chroma are the two biggest research-grade optical filter companies. They are both quite responsive to making custom filters for your specific application.

Research-grade cameras: Andor, Princeton Instruments, and Hamamatsu are the common big vendors. Hamamatsu has the biggest catalog of photomultiplier tubes. There are a number of smaller companies that periodically try to break into the market. Most camera vendors will let you have a camera for a couple weeks in your lab to test before you commit to buying it.

FLIR is the dominant vendor for infrared/thermal imaging and cameras.

If you need a small, easy-to-use spectrometer, Ocean Optics is a good source. If you need a bigger spectrometer with a scanning grating, Horiba, McPherson, Newport, and Princeton Instruments are some of the big main vendors.

Microscopes and objective lenses: There are four big companies competing in the research-grade microscope market. They are, alphabetically, Leica, Nikon, Olympus, and Zeiss.

Ultrafast lasers: The two largest companies making ultrafast lasers are Coherent and Newport. Other companies are breaking into the market as various patents start to expire.

Cryostats: Janus Research Company, Oxford Instruments, and Montana Instruments all make reputable cryostat systems.

3 Spectroscopy: So Many Squiggly Lines!

What is Optical Spectroscopy?

Optical spectroscopy is something that we all do every day. Spectroscopy, in simple terms, involves shining light onto an object and observing the change in the light after it interacts with that object. The change in the light after it interacts with the sample contains information about the sample. For example, when you go to the grocery store, the bananas are illuminated with white light. We use a wavelength sensitive detector, in this case our eyes, to measure wavelength-dependent reflectivity of the bananas to determine which bananas we would like to purchase versus which bananas are too ripe.

What Information Does Optical Spectroscopy Provide?

In the example of the bananas at the grocery store, what optical spectroscopy really provided was information about the energy levels of the molecules in the banana skins. There is a change in the energy levels of the molecules in the banana skin as the fruit undergoes the chemical reactions of ripening. This change in electronic energy levels of the molecules in the banana's skin is what we actually measured with this simple version of optical spectroscopy. We use the extra information that yellow bananas are ripe and ready to eat to make a decision about which bananas to buy.

Energy Levels

Molecules have electronic energy levels and vibrational energy levels (Section 1.2 and Table 1.3). We can use optical spectroscopy to better understand those energy levels.

The electronic energy levels are a little easier to think about first. If we consider the Bohr model of the atom, there are discrete stable orbitals in which electrons can be found orbiting the atomic nucleus. A photon, as previously discussed, has a certain discrete amount of energy. If the energy of a photon matches the difference in energy between two electronic levels in an atom, then the electron can absorb the energy of the photon and jump up from the lower electronic energy level to the higher electronic energy level. If the energy of the photon is less than the difference in energy levels, then the photon won't be absorbed and will pass through the material.

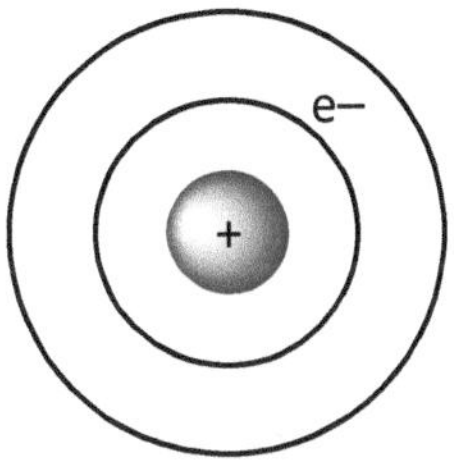

Figure 3.1

Bohr model of an atom depicting different electronic energy levels.

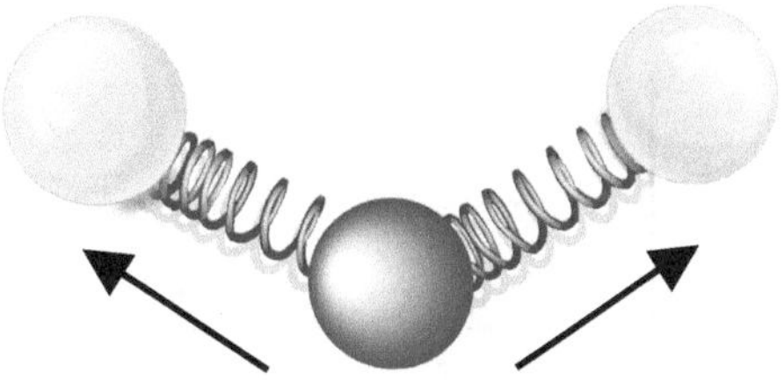

Figure 3.2

Symmetric stretch of a water molecule, showing vibrational motion of the molecule.

To return to the lower electronic energy level, the electron must get rid of the extra energy (due to the law of conservation of energy), which it often does by emitting a photon. The photon emitted from the electrons downward will always have a very specific energy (wavelength) that contains information about exactly which two energy levels the electron transitioned between. Atoms and molecules have very specific sets of energy levels, and by monitoring the light that is absorbed or emitted as electrons transition between these energy levels, chemical information about the sample can be deduced.

Electronic energy levels can be probed by UV-VIS absorption spectroscopy, fluorescence spectroscopy, and photoluminescence spectroscopy. Time-resolved spectroscopy techniques such as fluorescence lifetime imaging microscopy (FLIM) or time-resolved photoluminescence (TRPL) will measure how long an electron stays in an excited state, which will provide information about a molecule's local chemical environment.

To think about vibrational energy levels, visualize the classical model of a molecule as a system of balls and springs. Different atoms correspond to balls of different mass. Different bonds are springs with different stiffness. Each system of balls and springs has a certain set of resonant frequencies at which it will vibrate. Fourier transform infrared (FTIR) spectroscopy and Raman spectroscopy probe differences in vibrational energy levels.

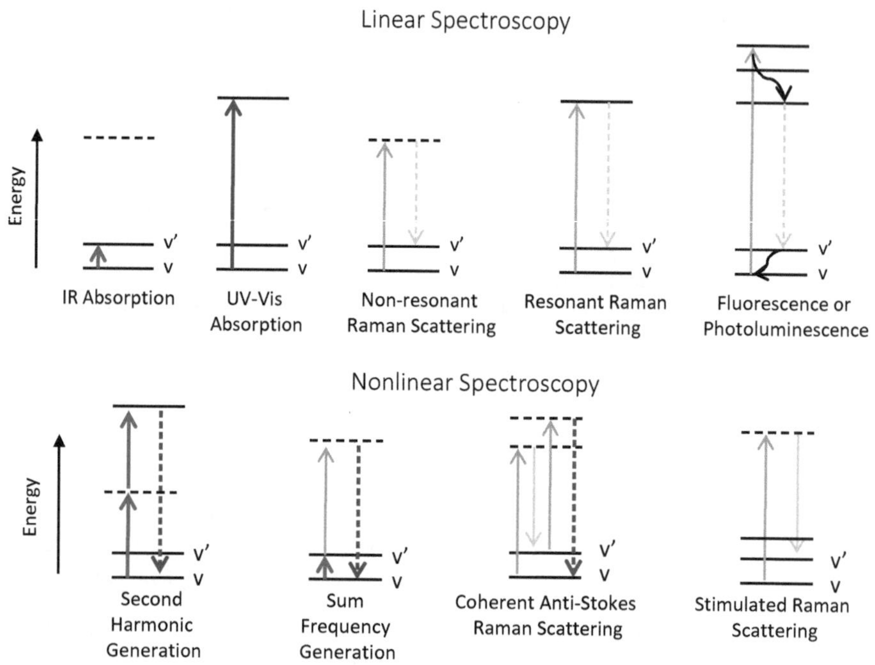

Figure 3.3

A family portrait of the energy level diagrams for some common spectroscopy techniques. Solid arrows denote excitation light. Dotted arrows denote emitted or scattered light. Solid horizontal lines are real energy levels. Dotted horizontal lines are virtual energy levels, through which a molecule can pass but cannot stay. Energy level diagrams are technically called Jablonski diagrams.

Spectroscopic techniques can be divided into linear and nonlinear techniques. Roughly put, the linear techniques involve processes in which there is one photon in and one photon out. In nonlinear techniques, there are multiple photons in and one photon out. Nonlinear techniques are inherently more complicated to implement, but they can provide further important information about the sample.

Spectroscopy can also be split up as micro-spectroscopy and macro-spectroscopy. The main difference between the two is that in micro-spectroscopy, signals are recorded through a microscope objective, providing highly localized information. Macro-spectroscopy provides spatially averaged information about a bulk sample.

In this section, we will introduce a number of optical spectroscopy techniques, so that you may understand what information is provided by each technique. This will hopefully allow you to make an informed decision about which technique is best suited to answering your scientific question.

3.1 UV-VIS-NIR Absorption Spectroscopy

Commonly just called UV-VIS spectroscopy, UV-VIS-NIR absorption spectroscopy is an absorption spectroscopic technique using ultraviolet (UV), visible, and sometimes near infrared (IR) light (200–2500 nm if everything is included). UV-VIS spectroscopy asks: "What color of light was absorbed by the sample and how much light was absorbed by the sample?"

A bench-top UV-VIS spectrometer (sometimes referred to as a spectrophotometer) can often accommodate a specular reflection accessory or diffuse reflection accessory to additionally measure what color of light was reflected from the sample and how much light was reflected from the sample.

Underlying Physical Principles

UV-VIS is a measurement of the upward energy transitions. As discussed in the introduction to this chapter, when light hits a sample, a photon can be absorbed, causing an electronic transition in which an electron moves from one ground state energy level to a higher exited state energy level. These electronic transitions for most materials reside in the visible to ultraviolet energy range. UV-VIS absorption spectroscopy is usually performed by illuminating the sample successively with different light energies and measuring how much light is transmitted (or reflected) by the sample relative to a blank sample measurement.

Classical UV-VIS absorption spectroscopy is usually performed in transmission mode from 200–1100 nm, relying on an Si photodetector or photomultiplier tube to serve as the detector. Some systems with multiple detectors can operate further into the near IR spectral region, denoted UV-VIS-NIR. The near IR detector can be an indium gallium arsenide (InGaAs) or lead sulfide (PbS) based detector.

Some systems can take specular reflection measurements, diffuse reflectance, or diffuse transmission measurements. Extra accessories are needed for these measurements. Micro-UV-VIS absorption spectroscopy can be performed with a spectrometer attached to a microscope, but these systems are much less common. Typically, UV-VIS absorption spectroscopy is a macro-spectroscopic technique.

UV-VIS data can be plotted either as wavelength versus percent transmittance (%T), percent reflectance (%R), or as wavelength versus absorbance (Abs). The absorbance units are optical density (OD). It is important to note that percent transmittance is a linear scale, whereas absorbance is a log scale. In a %T plot, the information about the sample is a series of dips from 100 percent transmission. In an absorbance plot, the signal is a series of peaks from a base line of 0 absorbance. The advantage of an absorbance plot is that the peaks are linear with concentration of the absorbing molecule. This is described by the Beer–Lambert law, which is more fully explained in the next section.

When taking an absorption or reflection measurement, set up the software to record a baseline and a zero correction. The baseline measurement records the spectrum from a "blank" sample to determine how 100 percent transmission (or 100 percent reflection) would look in your experiment. Baseline correction will account and correct for the variation in intensity as a function of wavelength of your light source, the throughput on the diffraction grating, the quantum efficiency of the detector, etc. All of these factors together are sometimes referred to as the instrument response function (IRF). In transmission experiments, the blank could be air, or a cuvette with solvent and no solute, or a bare substrate. In reflection experiments, an object usually needs to be placed at the sample position to reflect light back to the detector. A polymer called Spectralon is often used as a diffuse reflectance reference material. In specular reflectance measurements, UV-enhanced aluminum mirrors are often used for reference samples. A very important point about reflectance measurements is that you are making a measurement relative to the reference material. Aluminum mirrors have a flat reflectance across a wide spectral range, but they only have 90 percent reflectance. If you then try to measure an object that is 95 percent reflective, for example, the instrument will tell you that your sample is 105 percent reflective. This of course makes no physical sense, until you realize that you made a reflectivity measurement relative to the aluminum mirror. If you want to make absolute measurements you will need a National Institute of Standards and Technology traceable reference sample, which should come with a digital file that will allow you to correct your data for absolute measurements.

The zero correction will ask you to block the light source and then will measure the dark counts of the detector. Dark counts are the thermally excited electrons, as opposed to photoexcited electrons in the detectors.

What Scientific Questions Can Be Asked?

If your question is about the concentration of a solute in a solvent, an absorbance plot is more useful as the peak heights will be linearly related to concentration in the sample. The depths of the dips in a %T plot are *not* linearly related to the concentration. This is explained by the Beer–Lambert law.

The Beer–Lambert law (also occasionally known in introductory chemistry courses as the "abc" law) states that the total absorbance (A) of a sample is equal to how much light each absorbing molecule in the sample absorbs, which is a fundamental constant known as the molecular absorptivity of the sample (ε, also sometimes symbolized a), times the path length over which the light interacts with the sample (l, sometimes also symbolized b) times the concentration of the absorbing molecules (c).

$$A = \varepsilon lc = \log_{10} \frac{I}{I_0} \tag{3.1}$$

The total absorbance of the sample can be measured with a UV-VIS absorption spectrometer by determining $\log_{10} I/I_0$, where I is the intensity of the light transmitted through the sample and I_0 is the intensity of light transmitted without a sample. A blank sample is run first to determine I_0. The blank can be a cuvette with the pure solvent in the case of a liquid sample, the substrate in the case of a thin film on a substrate, or in some cases just air. The sample is then inserted into the beam path and I is measured.

Measuring which wavelength of light is absorbed provides information about the electronic energy levels of the sample. This can be used to detect changes in molecular composition, for instance in a hemeprotein bound to an O_2 or a CO molecule. Often the peaks are broad and extra information about the sample is needed to interpret any peak shifts.

UV-VIS absorption spectroscopy can also be used to calculate concentration levels of a light-absorbing solute in a solution. A set of standards would be made and measured first, and then the unknown would be plotted against the standard calibration curve. For amorphous semiconductor films, a Tauc plot can be generated to determine the edge of the bandgap.

With reflection and transmission measurements as a function of angle and polarization, the index of refraction of a thin film can be calculated.

Strengths and Limitations

UV-VIS absorption spectroscopy is generally straightforward to perform and is applicable to a wide range of samples. Off-the-shelf and bench-top spectrometers can be purchased from a variety of vendors and on the used equipment market.

This technique's biggest weakness is that the electronic absorption bands of molecules tend to be fairly wide and often overlap with each other, leading to a lack of specificity, especially in inhomogeneous samples with multiple components.

The measurement itself records the amount of light transmitted relative to a blank sample. The results are often plotted in absorbance units as absorbance units are a log scale and can be easily correlated to concentrations. There is an assumption, though, that any light that is not transmitted through the sample was absorbed by the sample. This assumption breaks down if the sample scatters much light or reflects specific wavelengths, as is sometimes the case with nanostructured materials. Other measurement accessories such as an integrating sphere or a variable-angle specular reflection accessory may be needed in such cases to account for the light that is neither transmitted nor absorbed.

What Samples Are Appropriate?

Nearly any sample that you can get light through, or bounce light off of, is appropriate. Highly scattering samples won't work well, as discussed above, unless you are using an integrating sphere to capture the diffuse reflected light. Samples can be dissolved in a solvent and placed in a cuvette. Specialized gas cuvettes can also be purchased and placed in the beam path to analyze gas compositions.

Schematics

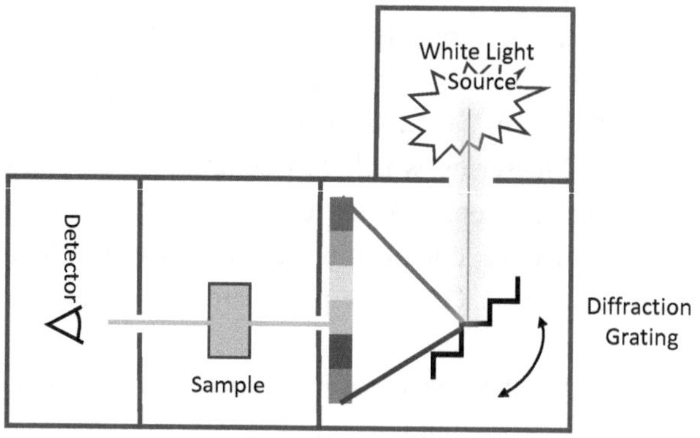

Figure 3.4

Schematic of UV-VIS spectrometer in transmission mode. The diffraction grating spreads out the white light like a rainbow and a slit selects out a specific color. The angle of the diffraction grating is scanned and the absorbance of the sample at each wavelength is detected.

Common Pitfalls

The quantum efficiency (QE) of the Si photodetector is low in the 900–1100 nm range compared to the detector's QE in the rest of the UV and visible range. Thus, there is a danger of misinterpreting noise as signal in this region. Systems that have detection at wavelengths longer than 1100 nm have a second detector.

There can also be stitching artifacts when the system switches between visible and near IR detectors, when it switches between the visible and near IR diffraction gratings, and when it switches from UV light sources such as a deuterium lamp to a visible and near IR light source such as a tungsten lamp. The exact wavelength where switches occur can be set in the software. If you suspect that there is a stitching artifact, try adjusting the switching point.

Using a sample with too high of a concentration will absorb too much light and an accurate measurement cannot be made.

A sample which is highly scattering such as a cloudy or turbid sample will not give you an accurate transmission measurement, as the light scattered out of the optical path is not taken into account. An integrating sphere would need to be used and diffuse reflectance and diffuse transmission measurements would need to be made to determine what wavelengths are truly absorbed by the sample.

A sample in which the solute is not yet completely dissolved will also not give accurate results since it may evolve in time. In general, samples that are dynamic and changing in time need to be treated carefully. The measurement does take a finite amount of time to perform. You may need to trade resolution and signal to noise for speed in these cases.

Some UV-VIS spectroscopy systems use two lamps, a deuterium lamp for the UV range and a tungsten lamp for the visible range. These lamps have a typical warm-up time of ~10–20 minutes and they have a finite lifetime (number of hours they can be on), so be sure to turn them off when you are not using them and to allow sufficient warm-up time when you turn them on. The deuterium also tends to diffuse out of the lamp over time, so periodically check the 656.1 nm deuterium emission line relative to the 656.22 hydrogen emission line to gauge the health of your deuterium lamp.

For liquid samples, rectangular cells called cuvettes are used to hold the sample. Disposable polystyrene cuvettes absorb strongly at wavelengths shorter than 350 nm. To work in the ultraviolet region, quartz cuvettes are needed. Micro-volume cuvettes often have a tapered region of the cuvette walls to reduce the amount of sample needed for analysis. Care should be taken with the orientation of the cuvettes when loading into the spectrophotometer so that the tapered region does not intersect the beam. The tapered regions can bend the light out of the collection optics path and you will get no signal.

As mentioned earlier, reflectance measurements are often relative measurements, not absolute measurements. Be aware of how reflective your reference reflective sample is.

Sample Data and Data Interpretation

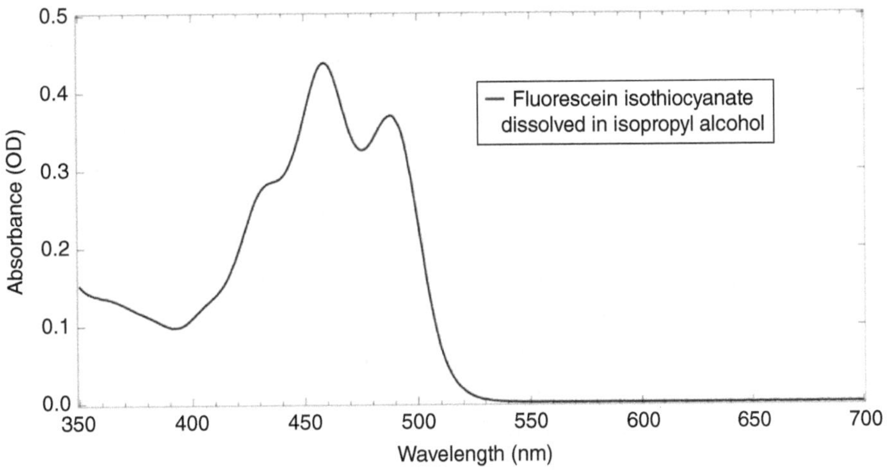

Figure 3.5

UV-VIS absorption spectrum of fluorescein isothiocyanate dissolved in isopropyl alcohol in a 1 cm path length polystyrene cuvette. The absorption peaks correspond to the differences in energy levels in the dye, or in other words the energies of the photons at which the fluorescein isothiocyanate absorbs photons efficiently.

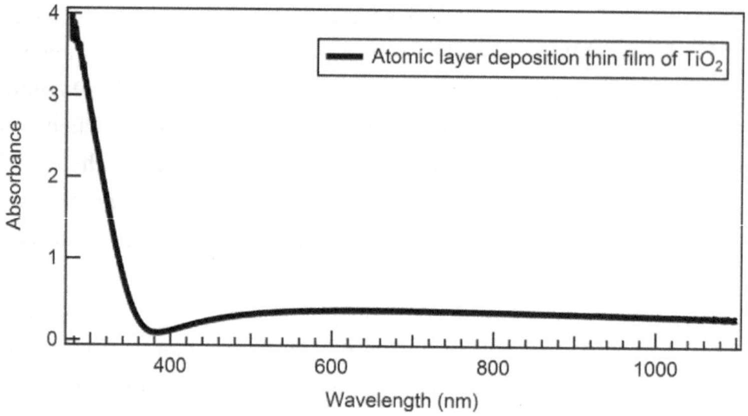

Figure 3.6

UV-VIS absorption spectrum of atomic layer deposition thin film of TiO_2 showing strong absorption as the energy of the excitation photons becomes larger than the bandgap of the material. Sample courtesy of Mac Hathaway.

3.2 Fluorescence Spectroscopy

Fluorescence spectroscopy is used to characterize the emission light from a sample after excitation with light. Fluorescent dyes and fluorescent proteins are used as labels in a number of microscopy techniques.

Underlying Physical Principles

Fluorescence spectroscopy measures the energy of the photon emitted during a downward radiative transition. In some ways, this is the complement of the UV-VIS absorption spectroscopy. Fluorescence and photoluminescence are the same physical phenomenon. Fluorescence is the term used by chemists and biologists and is usually applied to organic dyes and fluorescent proteins. Photoluminescence is the term used by physicists, material scientists, and electrical engineers, and is usually applied to solid state materials and will be discussed at length in the following section.

A spectrofluorometer usually consists of a white light source with a diffraction grating to allow for the selection of a specific excitation wavelength, a sample chamber, a diffraction grating to allow for the specific selection of an emission wavelength, and a cooled point detector or a cooled array detector.

A **fluorophore** is a molecule with fluorescent properties. Organic dyes that fluoresce typically have a conjugated ring structure that allows electrons in the excited state to become delocalized. As a general rule, larger ring structures have redder (lower energy) fluorescence emission.

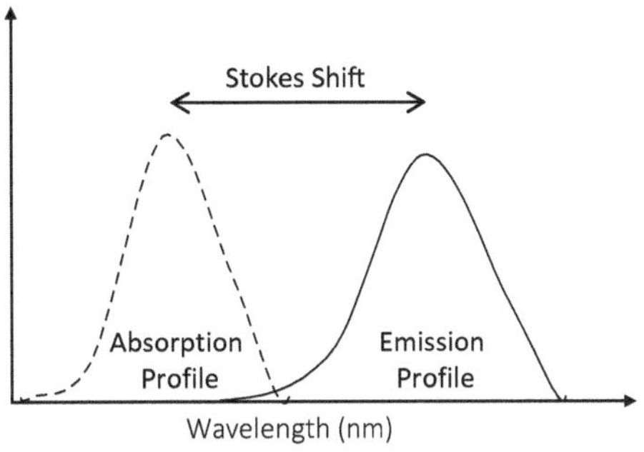

Figure 3.7

The Stokes shift is the difference between the excitation maximum and the emission maximum.

Quantum dots are pieces of semiconducting materials that are small enough to start exhibiting quantum confinement effects on the energy levels of the electrons. Similar to organic fluorophores, the larger the quantum dot, the lower energy the emission.

The difference in the peak position of the dyes absorbance and the peak of the dyes emission is referred to as the **Stokes shift**.

The **quantum yield (QY)** of a dye is a measure of the efficiency of the fluorophore, and is therefore defined as the ratio of the number of emitted photons to the number of excitation photons. The emitted photons leave the sample isotropically (in all directions), so to calculate the total number of emitted photons you will need to calculate the solid angle over which you are collecting the emission photons and scale that number of detected emitted photons appropriately to account for an entire sphere of emission. The number of excitation photons can be calculated from the power of the excitation beam measured with an optical power meter (Section 2.2.9).

In fluorescence spectroscopy, you typically collect the signal at 90 degrees to the angle of incidence, as this helps to spatially filter the excess excitation light that did not interact with your sample and thus does not carry the information about your sample that you are interested in.

When collecting a fluorescence spectrum, first record a background measurement to correct for the dark counts on the detector. The dark counts are thermally excited, as opposed to photoexcited electrons, and will yield a raised detector baseline on your measurement. For this reason, many detectors are cooled to minimize the thermal excitation contributing to background noise. It is important to wait for the temperature of the detector to stabilize before taking your background measurement, or else too large of a dark count contribution will be subtracted from your data and you will see a negative baseline.

If working with liquid samples, make sure that your cuvettes (square liquid sample cells) are optical grade and polished on all four sides. Note that cuvettes that are intended for UV-VIS measurements are only polished on two sides, so be sure to use the correct type. Similar to UV-VIS measurements, if you are working in the UV range (<350 nm), use a quartz cuvette since the disposable polystyrene cuvettes will absorb strongly below 350 nm.

Two types of scans can be performed with a spectrofluorometer. An excitation scan holds the emission spectrometer fixed and scans the excitation spectrometer to determine the excitation profile of the sample. An emission scan holds the excitation spectrometer fixed and scans the emission spectrometer to determine the emission profile.

The polarization of the excitation and emission light can be controlled to measure the fluorescence anisotropy, which is the degree to which the polarization state of the excitation light is preserved (or not) after interaction with the sample.

What Scientific Questions Can Be Asked?

Fluorescence is often used as a tool to mark some other process. For example, living and dead cells can be distinguished by labeling live or dead cells with a particular fluorescent dye. Alternatively, particular biomolecules (e.g., proteins or

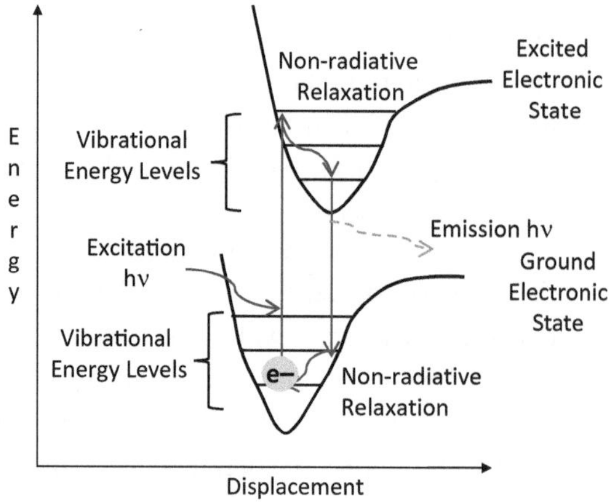

Figure 3.8

Energy level diagram for fluorescence (photoluminescence). A higher-energy photon is absorbed and causes an electron to transition from a ground electronic state to an excited electronic state. The electron typically undergoes non-radiative relaxation to the lowest portion of the excited electronic state energy manifold and then emits a lower-energy photon as it drops back down to the ground electronic state energy manifold.

antibodies or DNA) or analytes can be attached to fluorescent dyes, and the concentration can be measured using fluorescence spectroscopy (for instance to determine the concentration of a pollutant in drinking water). Fluorescence measurements versus time can also provide information about how a chemical reaction is proceeding.

Strengths and Limitations

Fluorescence can have a very low detection limit, in theory down to a single fluorescent molecule with a sensitive enough detector (e.g., a single photon counter or sensitive photomultiplier tube). Fluorescence is limited to molecules with strong, emissive electronic transitions. Additionally, solvents and other background materials that auto-fluoresce should be avoided, so that the signal from the analyte of interest can stand out against the background.

What Samples Are Appropriate?

Spectrofluorometers are usually configured to accept liquid samples in a square 1 cm cuvette.

Schematic

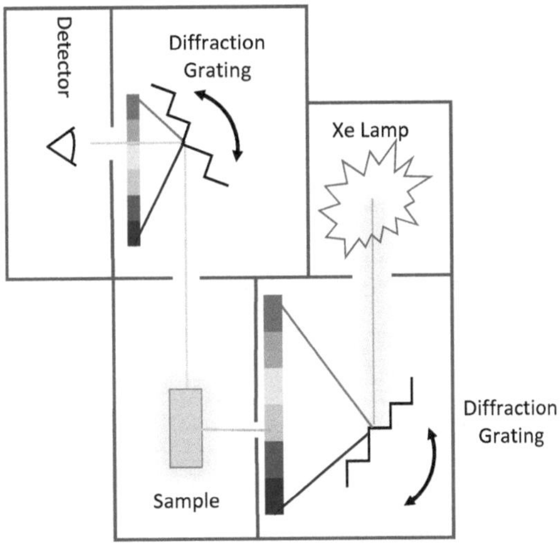

Figure 3.9

Schematic of a fluorescence spectrometer.

A **plate reader** is a fluorescence spectrometer that is designed to automatically read the fluorescence from a multi-well plate. A multi-well plate is a polystyrene plate with a number of cylindrical wells (6, 12, 24, 48, 96, or 384 wells typically). The multi-well plates allow you to easily perform a number of experiments in parallel. Plate readers are very common in biology labs, as different biological assays often use fluorescence reporters. Due to the automation needed to address the different well positions automatically, a plate reader should be purchased from vendors. You can also perform absorbance measurements on a plate reader, which involves using the plate reader as a UV-VIS spectrometer.

Common Pitfalls

Peaks from the second order of the excitation light on the diffraction grating can show up in your data and be very misleading. Be skeptical of any strong peaks that are almost exactly double your excitation light wavelength. In reality, these peaks appear in the data as strong peaks that are slightly less than double the excitation light wavelength.

Again, be aware of what materials your cuvettes are made out of and what wavelength range you are working in. Disposable polystyrene cuvettes are very convenient, but will not work for ultraviolet excitation. To work in the UV range you will need fused silica or quartz cuvettes. Additionally, avoid solvents that might dissolve polystyrene or have fluorescence stronger than your sample of interest.

Sample Data and Data Interpretation

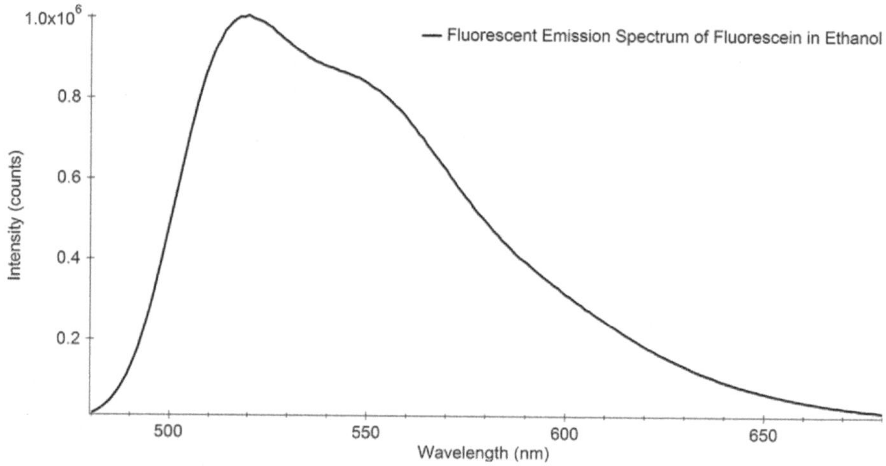

Figure 3.10

Sample fluorescence data.

3.3 Photoluminescence Spectroscopy (and an Early Introduction to Imaging)

Photoluminescence (PL) is fluorescence for physicists and material scientists. The underlying physics is really the same, but the language used to describe it is a little different and the types of samples studied tend to be a little different. We have included two separate sections to help the different scientific communities in their own terminology. Here, we discuss PL spectroscopy *and* imaging, and forego the discussion of PL imaging in Chapter 4 in the hopes of giving an early introduction to imaging, since the principles discussed in the imaging setups here are useful for understanding both the spectroscopy sections that follow throughout this chapter, as well as the imaging techniques in Chapter 4.

PL spectroscopy setups are often homemade, but a commercial Raman microscope can be used for micro-PL measurements and mapping by changing the units in which the resultant spectra are plotted in from cm^{-1} to nm or eV. For steady state bulk PL measurements, a spectrofluorometer can also be used.

Note that PL is another abbreviation used in the semiconductor industry and microelectronics for photolithography (Section 4.20); be sure you have the right interpretation of PL, depending on what you're reading and who you're talking with.

Underlying Physical Principles

When light hits a sample, the energy in the photon can be absorbed by an electron. The electron is then promoted from a "**ground state**" energy level to an "**excited state**" energy level. After being promoted to an excited state, the electron can "relax" to a lower energy level within the excited state energy **manifold** (set of energy levels) by transferring energy into lattice vibrations. Lattice vibrations are often thought of as "heat," so the electron is said to "**thermalize**" as it undergoes this process. Eventually, the electron will lose all the excess energy that it absorbed from the excitation photon and the electron will transition back to a ground state energy level.

The transition back to a ground state energy level can be a **non-radiative transition** or a **radiative transition**. If the transition is non-radiative, then the excess energy in the electron is transferred to lattice vibrations and no light is emitted from the sample. However, if the transition back to the ground state is a radiative transition, then the extra energy is emitted as a photon. The energy of the photon (color of the light) that is emitted tells you the energy difference between excited state and ground state energy levels that the electron transitioned between.

The energy of the emitted photon will always be at a lower energy (longer wavelength) than the excitation photon.

The light that is emitted from the sample contains information about the energy levels of the sample. Photoluminescence refers to light emitted from a sample after excitation with a photon. There are many other ways to excite a sample and generate luminescence. For instance, cathodoluminescence (CL) refers to light emitted from a sample after excitation with an electron beam. In this section we will focus on just PL spectroscopy, although many of the concepts can be extended to other types of excitation.

The terms photoluminescence and fluorescence are really describing the same physical process. By convention, the term photoluminescence is used when talking about solid state systems such as a semiconductor sample. The term fluorescence is used when talking about organic molecules such as dyes or light emitting proteins.

Phosphorescence is a slightly different physical process. In phosphorescence the excited electron ends up stuck in a "**triplet state.**" In the triplet state, the spin of the excited electron has become flipped while in the excited state. Due to the **Pauli exclusion principle**, the spin of the electron must be flipped back before it can return to the ground state. In some materials, the probability for the electron spin to be flipped back is low, which means that electrons can remain in the excited state for minutes, hours, or even days. Phosphorescence is what enables glow-in-the-dark stickers to continue glowing for such a long time.

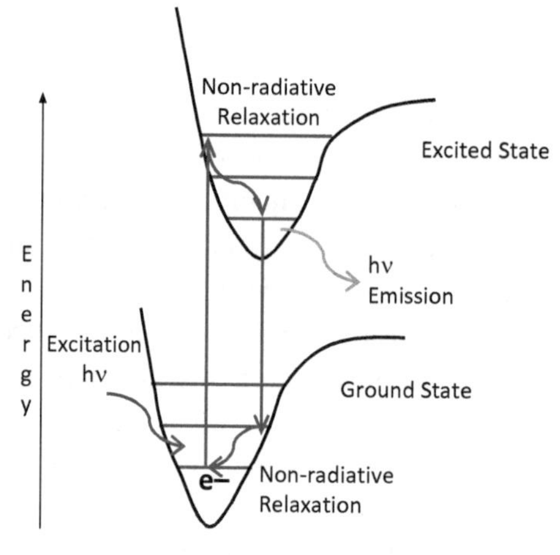

Figure 3.11

Energy diagram of photoluminescence. The emission photon is always at a lower energy (longer wavelength) than the excitation photon.

What Scientific Questions Can Be Asked?

In this section, we are going to focus on the solid state and semiconductor samples that are traditionally probed by "photoluminescence." We discussed samples traditionally probed by "fluorescence" spectroscopy in the previous section.

Photoluminescence is predominantly used to characterize the relative energies of electronic energy levels in a solid state material. There are many materials in which the energy levels can be carefully engineered and the exact energy levels are important to the use and performance of the material.

For instance, the **bandgap** (i.e., the difference in energy between the conduction and valence bands) of semiconductors or quantum wells (such as those used for LEDs or lasers – Section 2.2) can be measured using PL spectroscopy. Moreover, PL spectroscopy can characterize "defect states," "trap states," "doping levels," and "surface states," which refer to electronic energy levels within the bandgap which are caused by defects or impurities in a material.

Time-resolved PL spectroscopy can give you a picture of the ultrafast electronic dynamics (carrier lifetime) of a sample. A fast pulsed laser and a fast detector are needed to do such experiments. We will reserve our discussion of time-resolved techniques for Section 3.8. Time-resolved PL spectroscopy is common for characterizing carrier lifetimes in semiconductor samples in the context of photonic devices such as solar cells, quantum dots, LEDs, CCDs, etc.

Photoluminescence mapping or imaging can tell you about the spatial location of defects or dopants in a sample. Although this is the spectroscopy section, some discussion of PL imaging will be included here, since white light imaging is usually extremely helpful to find the sample feature from which you record PL spectra and thus there is a lot of overlap of PL imaging and PL spectroscopy.

Quantum yield refers to the percentage of photons absorbed which yield an emitted photon versus those which undergo non-radiative transitions. In all cases, $QY \leq 1$. Materials with PL $QY > \sim 0.05$ (only 5 percent of absorbed photons result in emitted photons) can still display very bright luminescence. In some samples in which it is beneficial to emit lots of light (like a material for an LED application), you would like to have a high quantum yield. However, in samples in which emitting light is a detriment (such as materials for a solar cell), you would like to have a low quantum yield.

Strengths and Limitations

For PL spectroscopy to be possible, a material must relax though a radiative transition. Some materials are much more likely to relax through a radiative transition than others. "**Direct bandgap**" materials, such as indium phosphide, are much more likely to relax through a radiative transition and thus give strong PL signal. "**Indirect bandgap**" materials, such as silicon, are much more likely to relax through a non-radiative transition and thus have a very weak PL signal.

Cooling the samples with weak PL signals down to cryogenic temperatures will enhance weak PL signals by "freezing out" some of the non-radiative relaxation pathways (since the non-radiative relaxation pathways rely on lattice vibrations and low temperatures inhibit atomic motion in crystals.) An optical cryostat is necessary for such experiments. Liquid nitrogen will allow you to cool the sample to 77 K. Liquid helium will allow you to cool to 4 K. A closed cycle cryostat is usually a good investment since liquid helium is becoming increasingly expensive and in short supply.

Materials with relatively lower energy ($<\sim 1.2$ eV) PL will require a sensitive IR detector such as a cooled InGaAs array or cooled mercury cadmium telluride (HgCdTe; MCT) detector.

Trying to characterize the energy levels in a small structure (single nanowire, single quantum well, etc.) will require a microscope setup to control the area of excitation and area of signal collection.

The PL signal of the conduction band to valence band transition is usually a very clear and strong peak that is easy to interpret. The smaller peaks from the different defects can be difficult to interpret and usually need other information about the sample – such as how it was prepared, and its composition as well as information from ultrafast spectroscopy and other literature or mathematical modeling – to assign correctly.

What Samples Are Appropriate?

While almost any sample can be probed with PL spectroscopy in general, the specifics of the setup available to you will limit which samples are appropriate. Some things to consider are:

- Is there an appropriate excitation source for your sample? The energy of the excitation photon needs to be larger than the energy of the transition that you wish to probe.
- Over what range is the detector sensitive? If you are interested in probing transitions of less than 1.2 eV you will need a sensitive IR detector, either an InGaAs or MCT detector.
- Will you need to cool your sample? If you have an indirect bandgap material, you will probably have to cool the sample to get a good PL signal.
- Will you need a microscope setup? Are you interested in only probing a specific small structure on the sample or is probing a larger area okay?

Schematics

Photoluminescence spectroscopy can be configured for a bulk or micro measurement. Bulk PL records a spectrum from a large area (as big as the diameter of your excitation laser beam) of the excited sample, whereas micro-PL (μ-PL) spectroscopy allows excitation and recording of PL spectra from an area of sample as small as ~1 μm, depending on the wavelength of your excitation laser. Thus, bulk PL is appropriate for studying homogeneous samples with relatively large areas, such as uniform thin films or dispersions of nanoparticles on wafers or other substrates. μ-PL spectroscopy can be used to characterize the energy levels of inclusions or features in a larger sample, such as a particular nanostructure or quantum dot.

Photoluminescence imaging tells you spatially (in x, y, and potentially in z) where the emission is coming from and how the intensity of emission varies as a function of position in your sample. Imaging can be configured in two ways, analogous to spectroscopy: widefield excitation or μ-excitation PL imaging.

In **widefield excitation** PL imaging, a wide area of the sample is illuminated with the excitation wavelength and the emission from the entire area is subsequently imaged with an optical filter onto a camera. Thus, this technique might be used to characterize where the active region of a device is, or where quantum wells are located in a sample.

With μ-excitation, spatial information about the sample can also be collected, usually through PL mapping. Photoluminescence mapping can be achieved either by raster scanning a small laser spot across an area of the sample (see, e.g., Section 4.9) with a set of scanning mirrors or by moving the sample relative to the fixed laser spot position with a motorized stage.

Spectroscopy and microscopy can be combined to yield "**hyperspectral**" imaging, in which each pixel in an image contains an additional axis of data related to the emission wavelengths of light coming from that area of a sample. In this case, the setup is configured similar to scanning PL imaging above, except, as the laser reaches each pixel, a PL spectrum is also recorded. You can end up with four- and five-dimensional data sets, such as (x position, y position, z position, spectrum, polarization, time, etc.). Hyperspectral imaging will be discussed in more detail in Section 3.7.

A good way to think about PL setups is to keep in mind that they have two optical pathways: the excitation pathway and the emission or collection pathway. In many setups, there is overlap between the two optical paths. There are certainly commercial setups that can be used for PL spectroscopy, but the general layout will be the same and it is useful to think about the layout to understand the limitations of a specific setup.

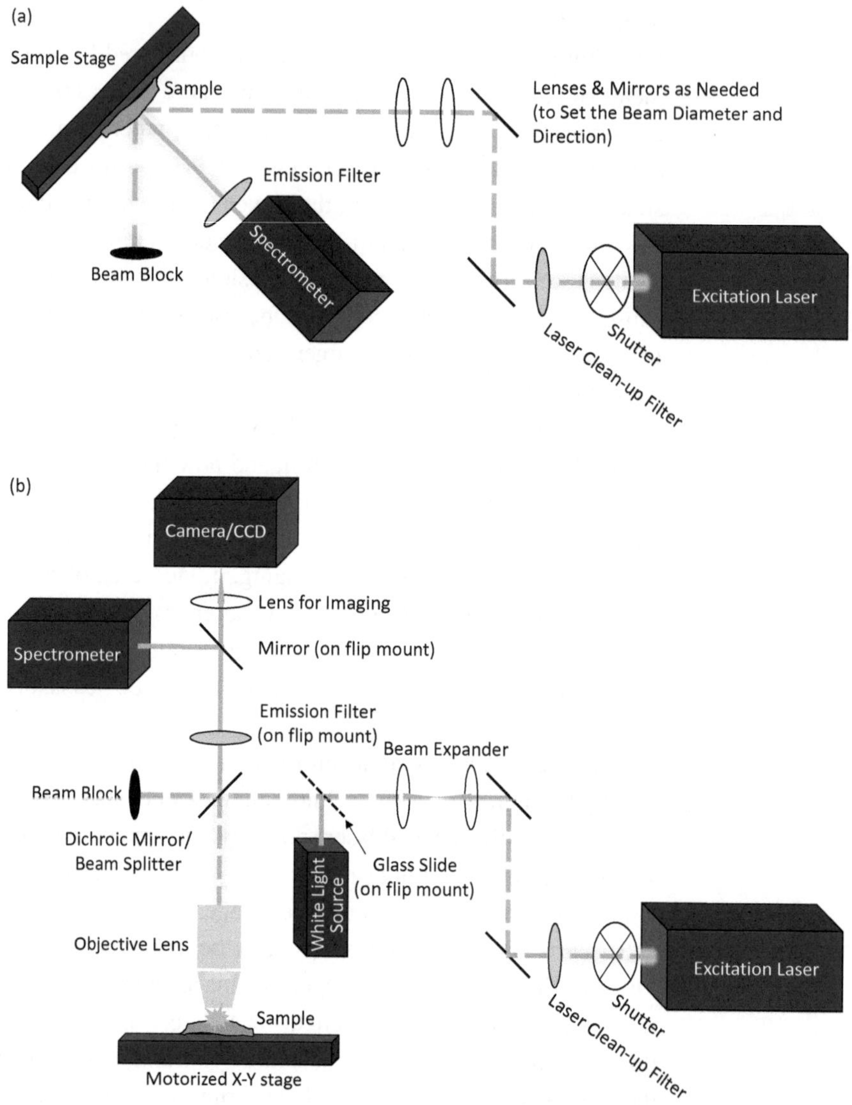

Figure 3.12

Schematics for the various configurations of PL imaging and spectroscopy. Dotted lines correspond to excitation light, whereas solid lines correspond to emitted light from a sample (or a plasma line from the laser). The gray rays correspond to broadband white light.

(a) Bulk PL spectroscopy. Note that the reflected excitation laser light is "dumped" out to the side, and the PL is collected at an angle not equal to the angle of incidence of the laser beam on the sample, leading to a cleaner signal.

(b) μ-PL spectroscopy and μ-PL imaging. It is possible to switch between imaging and spectroscopy simply by adjusting the flip mirror in the collection pathway. Note that for spectrometer/camera combinations which can perform both imaging and spectroscopy, the flip mirror is not necessary.

(c) Widefield PL imaging. Note that the laser beam size in the field of view can be adjusted by moving the lens along the beam path (via, e.g., a rail system) behind the objective lens.

(c)

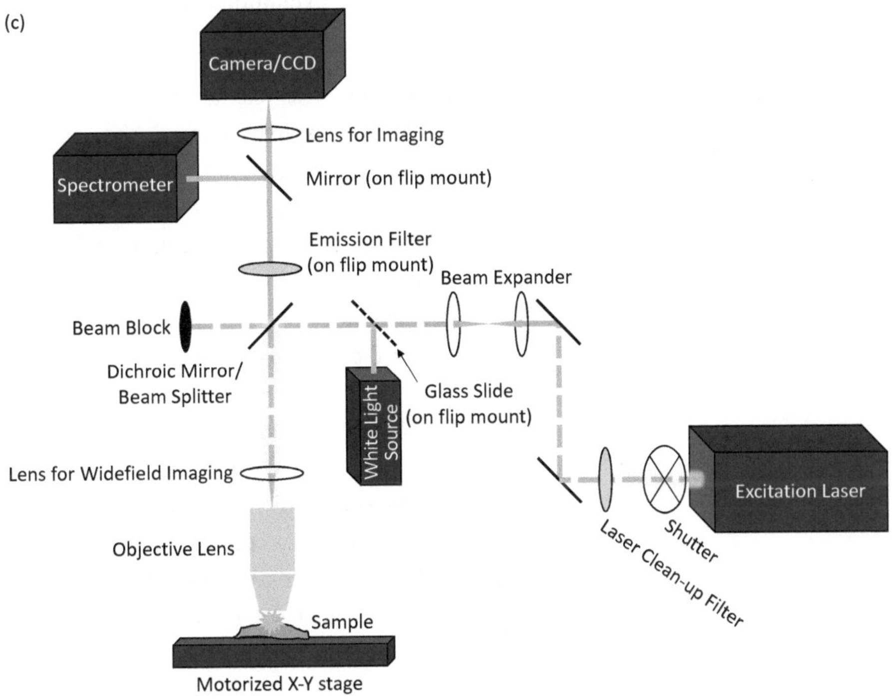

Figure 3.12 *(cont.)*

The excitation pathway begins with high-energy (short-wavelength) light, typically from a laser, which passes through a short-pass or bandpass filter to spectrally "clean up" the light traveling to the sample. A short-pass or bandpass filter (or perhaps multiple) is especially necessary for lasers with multiple lines (e.g., Ar^+ or Kr^+ ion lasers) since these lasers often have plasma emission lines that can be similar in energy to the emitted light from the sample you are trying to study. (For instance, Ar^+ ion lasers have a line at 514.5 nm which overlaps almost exactly with ~512 nm [2.42 eV] PL emission from cadmium sulfide [CdS], a semiconductor with emission in the green range.) An alternative to the use of a laser and filter would be a broadband white light source passed through a short-pass filter. However, the power densities achievable might be lower with a white light source than with a laser.

The excitation light is then reflected off of at least two mirrors that are used to align the beam (see Section 5.2 on why using at least two mirrors is important).

Two lenses functioning as a beam expander can be used to adjust the beam size (Section 5.9) – this is not always necessary if your laser is already collimated at a useful beam diameter.

The next components of the excitation pathway depend on the configuration of PL in use:

- For bulk PL spectroscopy, the excitation light can be subsequently passed directly to the sample. The excitation area is the same as the beam spot size.
- For μ-PL spectroscopy, μ-PL imaging, or hyperspectral imaging, excitation light is then reflected at 45° from a beam splitter or dichroic mirror. If scanning is required, a set of scanning mirrors could sit between the beam splitter/dichroic and the objective lens. The final component of the excitation pathway is an objective lens. The laser beam diameter should equal the diameter of the back focal aperture of the objective lens (Section 2.3.25) to give the smallest spot possible.
- For widefield PL imaging, similar to the above, excitation light is reflected off of a dichroic or beam splitter. However, another lens is used to focus the beam to as small a spot as possible in the back focal plane of the objective lens. This results in illumination of the entire field of view of the objective lens with the excitation laser light. (If necessary, intermediate spot sizes can be achieved by changing the size of the laser in the back focal plane of the objective lens by, e.g., moving the last focusing lens forward and backward along the beam path.)

To minimize stray reflections, in all of these cases, use a beam block to block the laser light that passes straight through the beam splitter/dichroic (or for bulk PL spectroscopy, to block the primary laser beam that reflects off of the sample).

Finally, we arrive at the sample under investigation. Samples for PL can be almost anything: solid, liquid, gas. These samples can sit in a cryostat for temperature-dependent measurements or in a cuvette for liquids or gases, as long as the windows of the cryostat or gas/liquid cells are transparent to the relevant wavelengths of light, and – in the cases where a microscope objective lens is used – the working distance of your objective lens is long enough to allow a focused spot to reach the sample through the sample holder window.

The collection pathway for the various PL techniques is configured as follows:

- For bulk PL spectroscopy, the emission passes through a long-pass filter to block the excitation light and then on to a spectrometer. Also, in most samples (especially those which would be appropriate for bulk measurements), PL is emitted in all directions, so we recommend that the collection angle is not the same as the angle of reflection of the excitation light, in order to minimize the amount of excitation laser light entering the detector.
- For μ-PL spectroscopy, emission is collected by the same objective lens, passes through the dichroic mirror, through another long-pass filter to exclude any remaining excitation light, and into a spectrometer.

- For μ- or widefield PL imaging, emission is collected by the same objective lens, passes through the dichroic mirror, through another long-pass filter to exclude any remaining excitation light, and then is focused back into the image plane (Section 2.3.21) by a lens before entering a camera.

The CCD or detector used should have sensitivity in the range of the emission wavelength of the material you are studying. For imaging and μ-PL measurements, we recommend mounting the sample, the objective lens, and the final focusing lens on micromanipulators (Section 2.3.10), since this will allow fine-tuning of the positions of the sample relative to the laser beam and enable easier fine alignment of all of the components. A standard microscope could also be used.

The choice of dichroic mirror and filters are very important for PL. The cutoffs for the short-pass filter (to clean up the laser), as well as the dichroic mirror and the long-pass filter (to exclude excitation light from the detector), must be between the excitation and emission wavelengths of interest. Steep cutoffs are better if the excitation and emission wavelengths are close to each other – that way, you won't clip the edge of the emission spectrum with the filter. Also, for dimmer samples (e.g., single nanoparticles or single-dye molecules), high OD/low transmission at the excitation wavelength and low OD/high transmission at the emission wavelength is important for the dichroic and the long-pass filter, since the laser light is usually very strong relative to the PL emission from such a small sample. Similarly, if the laser has plasma lines, a high-quality clean-up/short-pass filter is very important; without it, laser light could show up in the final spectrum and be confused for emission from the sample.

In general, you can switch back and forth between imaging and spectroscopy with a flip mirror or by using a beam splitter. Some monochromator/CCD combinations allow internal switching between a mirror for imaging and a grating for spectroscopy, so an additional beam path is not required. This is a strategy behind hyperspectral imaging – however, we recommend automating this process with some sort of computer control, especially if scanning is performed by an electronic stage or scanning mirrors, since otherwise you will have to flip the mirror back and forth a lot.

TIP: Don't fry your camera with the laser beam. The long-pass filter is essential for preventing the laser from getting to the detector and to exclude it from your measurement. But make sure this filter is in place (or the incident laser power is very, very low – check detector specs for permitted powers) before allowing the laser to reach your detector, or it may be damaged. If you must use a higher power laser for alignment, use an extra beam block before your detector just to be safe. Cameras cost enough that it is worth taking the extra precaution.

TIP: For all of the μ-PL spectroscopy setups, we recommend coupling a white light source into the beam path via glass slide, so that white light imaging can be performed in addition to PL imaging, as in Figure 3.12. (In this case, simply flip down or remove the long-pass filter for a broadband white light image of the sample before turning on the laser.) This is useful for finding an interesting area of sample and for figuring out what part of the sample you are looking at. See also Section 5.8.

TIP: Finding Focus on Your Sample

To get the sample into the proper focal plane of the objective lens, use a focused laser spot. Make sure the laser power is very low. Next, remove the long-pass filter (e.g., via flip/kinematic mount) to image the laser on the sample. Move the sample toward/away from the objective lens until the laser spot is as small as possible. This should be approximately optimal focus for PL spectroscopy and imaging (and white light imaging), too, although some chromatic aberration might slightly alter this since the wavelength of the excitation laser isn't exactly the same as that of the emission. Thus, some slight tweaking of the focus of the sample relative to the objective lens to maximize PL emission intensity and/or image focus might be necessary as a final optimization step. If the focused laser spot seems to move from side to side as you move in and out of focus, the laser beam is not going through the center of the microscope objective. You will need to walk the laser beam so that it is going down the center of the objective (see Section 5.1.2). The circular cross section of a laser beam going through the center of the microscope objective will become smaller or larger as the beam is focused and defocused, but the position of the beam will not be translated side to side or up and down.

TIP: Alignment of the Excitation Laser

To check the alignment of the laser through the objective lens, make sure the laser power is very low and image the laser spot on the sample and focus the laser on a very flat sample (e.g., glass slide, mirror, Si wafer, etc.). Then, move the flat sample out of focus (toward defocus, slightly away from the objective lens) such that the laser spot is not condensed anymore to a tiny spot. You will see sequential rings of alternating intensity – these are due to interference as the laser beam hits the sample and then rebounds back into the objective lens. Use the two alignment mirrors in the excitation pathway to make these disks as round and circularly symmetric as possible; the intensity should be even all the way around the rings. Then, adjust the emission pathway to center the laser spot in the center of the final lens and finally the camera. Iterate this process several times such that the defocused laser rings are circularly symmetric and round.

TIP: Transparent Samples

Some samples – in particular glass and quartz slides – are transparent. Similar to standard white light microscopy, you can focus the objective lens on both the top and bottom planes of these types of substrates. This can be confusing for PL imaging and spectroscopy, too, so double-check that your objective lens focal plane is in the proper plane of the front (not the back) of the sample.

Common Pitfalls

Best practices for collecting PL data include:

- Before characterizing your samples with PL spectroscopy, collect three "dummy" spectra. First, turn off the laser/other excitation source and collect a spectrum – this will tell you about background spectra from other light sources in the room. Common background spectra from lab objects include computer monitors and fluorescent lights. Second, take a "background" spectrum with a mirror in your sample position. The mirror should be close to 100 percent reflective for your laser wavelength – this will tell you how well your clean-up and laser exclusion filters are working. If you see laser plasma lines, or the primary laser peak, in your spectrum, purchase better (or more) filters and add them to the beam path. Lastly, take a PL spectrum of your substrate (e.g., glass slide or wafer) or sample holder (e.g., gas or liquid cuvette) without the sample on/in it – this will tell you about any contributions the substrate is making to the ultimate spectrum you record. Similarly, make sure that the substrate on which the sample sits does not photoluminesce at a wavelength relevant to the sample you are trying to study (e.g., an InP nanostructure emitting at ~920 nm on a GaAs wafer emitting at 870 nm).

- PL is highly dependent on incident laser excitation power. In fact, many organic samples and dyes will **photobleach** if the laser power is too high, which means that the light used to excite them actually rearranges bonds, rendering the dyes permanently unable to fluoresce. In general, we suggest longer exposure times with several integrations at low laser excitation powers instead of higher laser powers if the signal has low intensity because higher laser powers can cause artifacts in the sample due to heating or damage (especially for more "delicate" samples such as nanostructures). Always measure the laser power with a power meter at the sample location. We recommend performing a power-dependent measurement using some swappable broadband ND filters in the excitation pathway to characterize the excitation power-dependent luminescence properties of your samples.
- Guidelines for excitation power densities for some samples to avoid artifacts due to heating and damage are as follows. We assume a ~1 μm diameter focused spot, as in the case of μ-PL imaging or spectroscopy:
  - nanostructures, organics, biological, and other very delicate samples: <~100 μW;
  - robust thin films, nanostructures, and biological samples: <5 mW;
  - bulk crystals, wafers, and other very robust samples: <100 mW.
  - Ultimately though, some trial and error will be needed. It's likely you will blow up a few samples before you find an excitation power that works for you, and the appropriate power level depends on the relative response of your sample to the excitation wavelength you are using.
  - In general, the lower the excitation power, the better. This will preclude your ultimate reviewers from asking about sample heating or damage effects.
  - Another experiment to probe this effect is to take multiple spectra sequentially – if the spectrum from the same area of sample changes with time/the number of exposures, the sample is probably being damaged by the laser, or undergoing some change while exposed to the air or the surrounding atmosphere in its holder.
- Don't forget to turn off the white light and flip the long-pass filters back into the beam path when you record PL spectra and images. There is nothing more frustrating than waiting for a long exposure only to realize you left the white light on and it's drowning out the sample's luminescence.
- Some PL recording programs (such as those that control CCDs and gratings) allow you to stitch together spectra to get a spectrum over a wider range of wavelengths than what would be available in a single exposure without moving a grating. Be sure that the stitch is not in the center of the spectral

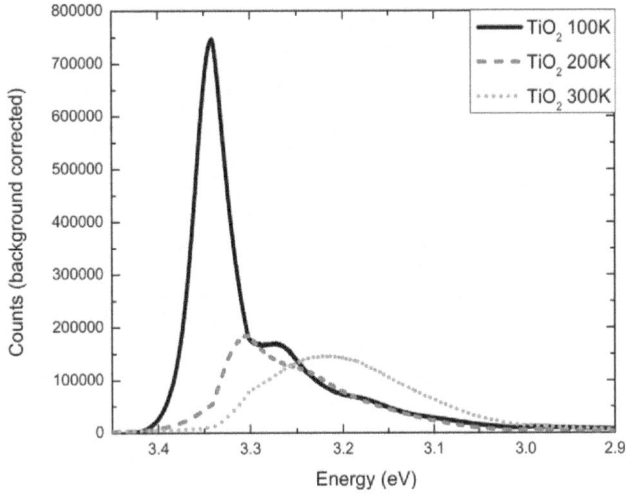

Figure 3.13

Temperature-dependent PL spectra of titanium dioxide films grown by atomic layer deposition on a silicon substrate. Note that lower temperatures tend to increase PL signal and to sharpen PL peaks. The largest peak (in this case around 3.3 eV) is usually the band to band transition and the smaller peaks at lower energies (like the small peaks at 1.7 eV) are defect states. The peak center is usually reported as the energy of the bandgap. Samples courtesy of Malcolm Hathaway.

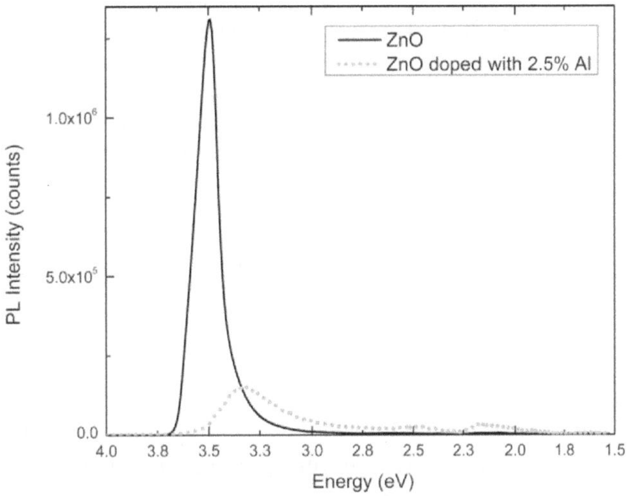

Figure 3.14

Room temperature PL spectra of zinc oxide (ZnO) film and aluminum-doped zinc oxide (AZO) films deposited by atomic layer deposition. You can see how the band to band transition (as indicated by the most intense PL peak) shifts due to the difference in bandgap for the two samples with different compositions. Also, the intensity of the lower energy peaks increases for the AZO samples, indicating that there are more defects in this sample compared to the ZnO film. Further sample processing, such as annealing, could be employed to try to decrease the number of defect states in the sample, using the PL signal as a way to monitor the number of defects. Samples courtesy of Malcolm Hathaway.

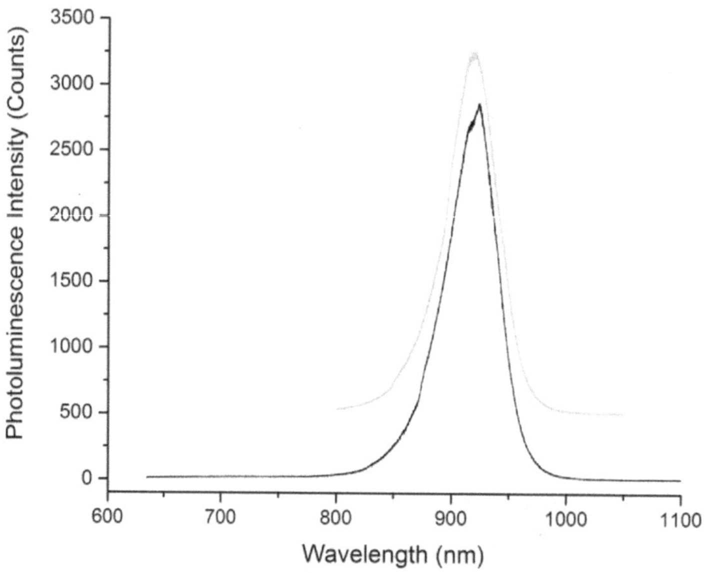

Figure 3.15

Two PL spectra from an indium phosphide (InP) wafer. A 633 nm laser was focused to a ~1.5 μm spot through a 50× magnification objective lens to excite the sample with an excitation power of ~65 μW. The long-pass/excitation filter we used had a cutoff at ~640 nm. The black line is our initial attempt at recording a spectrum; the gray is our second; the gray spectrum is offset from the baseline by 500 counts for clarity. As you can see, there is a sharp emission peak at ~922 nm, which is consistent with radiative recombination via the direct bandgap of InP at 1.344 eV at room temperature. However, the black spectrum was stitched together with multiple exposures by moving the grating with software control; in this case, two of the wavelength ranges for exposure overlapped right at ~920 nm, hence the bump in intensity at the peak of the black spectrum. We changed the wavelength range used for exposure for the second gray spectrum and this artifact disappeared. However, we note that the stitch point just shifted away, to ~850 nm, as you can see from the bump in the spectrum. So, it often helps to record several spectra over a few different ranges to isolate what is an artifact in the data versus what is real. For instance, some materials might have two PL peaks next to each other due to a trap state or some other defect.

feature of interest for your sample; otherwise, background intensity matching in different wavelength regions might give you unexpected artifacts in your spectrum (see Figure 3.15).

Sample Data and Interpretation

Photoluminescence spectra are plotted with intensity counts (number of emission photons detected) on the y-axis and emission photon energy (in nm or eV) on the x-axis. Sometimes, two x-axes are shown at the top and bottom of the graph; one plotted in nm and the other in eV.

3.4 Fourier Transform Infrared Spectroscopy

Fourier transform infrared (FTIR) spectroscopy is a vibrational spectroscopy technique. Vibrational spectroscopy techniques provide molecular information about the sample. Molecular information about a sample is often very important. For example, diamond and graphite are both pure carbon but their molecular arrangement is different, which leads to dramatically different physical properties. Diamonds are optically clear and hard while graphite is optically opaque and is soft. Materials with the same atomic composition but different molecular arrangements or different crystal structure are called **polymorphs**. FTIR and Raman (Section 3.5) spectroscopic techniques are sensitive to polymorphism in a material. Due to the automation needed for FTIR spectroscopy, systems are usually purchased from a vendor rather than homebuilt.

Underlying Physical Principle

A molecule can be crudely modeled as a classical mechanical system of balls and springs. Each molecule (or system of balls and springs) will have a unique set of resonant vibrational frequencies according to the simple harmonic oscillator (sometimes abbreviated SHO) model and thus a unique set of vibrational energy levels. These unique vibrational energy levels can be measured with FTIR and the molecular composition of the sample determined.

The simple harmonic oscillator model can be derived by combining Hooke's law:

$$F = -kx \tag{3.2}$$

and Newton's second law of motion:

$$F = ma \tag{3.3}$$

into

$$-kx = m\,\frac{d^2x}{dt^2} \tag{3.4}$$

Solving using the time-dependent solution:

$$x(t) = A\,\cos(\omega t + \varphi) \tag{3.5}$$

we find that

$$\omega = \sqrt{\frac{k}{m}} \tag{3.6}$$

So Equation 3.6 tells us that the resonant frequencies of the molecule (ω) depend on the mass of the atoms (m) in the molecule and on the strength of the

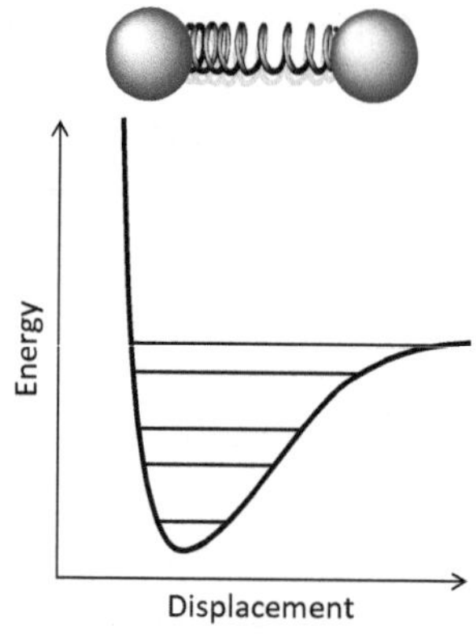

Figure 3.16

Vibrational energy level diagram showing a Morse potential energy well. Displacement is the distance between the atoms. As the atoms get too close, they start to repel each other, making the potential energy barrier rise sharply. As the atoms are stretched too far apart, the bond will eventually break. In FTIR, the molecule moves up a vibrational energy level by absorbing an infrared photon that has the exact same energy as the difference between the initial and the final energy levels.

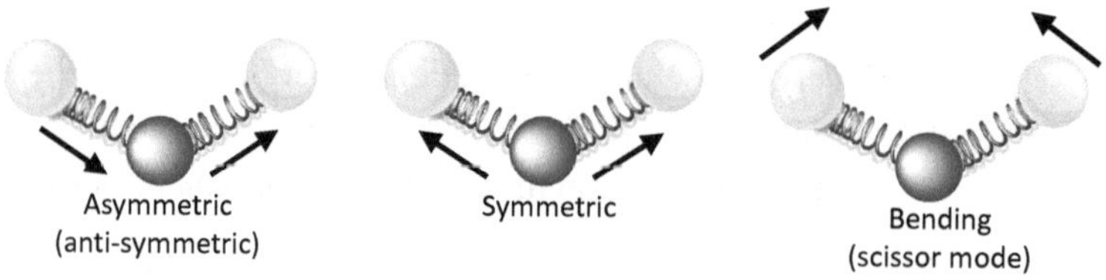

Figure 3.17

Normal modes of water. Antisymmetric mode, symmetric mode, and bending mode.

chemical bond (k) connecting them. It turns out that these frequencies are in resonance with the frequency of the electric field oscillations of light in the mid-IR region. So we can probe the vibrational energy level differences using light in the mid-IR region, 2.5–20 μm.

Different geometries of molecules can undergo different vibrational motions. The sets of vibrational motion that a molecule can undergo are called normal modes and can be described mathematically with group theory. A nonlinear molecule (such as water) will have $3N - 6$ normal modes, where N is the number of atoms in the molecule. So water has three normal modes: the antisymmetric mode (sometimes called the asymmetric mode), the symmetric mode, and the bending mode.

To determine the set of vibrational energy levels, an IR absorption spectrum is collected taking advantage of the fact that the probability that a molecule will absorb an IR photon is high when the energy of the photon exactly matches the energy difference between two vibrational energy levels.

Not all vibrational motions are IR active. For a molecule to be allowed to absorb an infrared photon, there must be a change in the net dipole moment with the molecular motion. Derivation of the selection rule is beyond the scope of this book. Water has a very strong infrared absorbance as it has a large change in its dipole moment with its three normal modes.

The original IR absorption spectrometer used a prism to separate the wavelengths of light spatially and measure the absorption of the sample one wavelength at a time. This was a slow process, and finding an appropriate material for the prism that works over the whole 2–20 μm mid-IR spectral range is almost impossible. In contrast to the dispersive approach, the FTIR spectrometer uses a broadband light source (all wavelengths simultaneously) and a Michelson interferometer to generate an interferogram that can be converted into an absorbance spectrum through a mathematical process called a Fourier transform.

A Michelson interferometer splits the light into two optical paths with a beam splitter. One optical path has a fixed mirror and the other optical path has a moving mirror. The mirrors send the light back to the beam splitter, where they can interfere constructively or destructively. When the two optical path lengths are exactly the same, then all the wavelengths of light in the two beams interfere constructively. When the moving mirror scans back and forth, different wavelengths interfere constructively or destructively, depending on the exact path length differences between the two arms of the Michelson interferometer. The recombined light beam is usually sent to a point detector, usually either a deuterated triglycerine sulfate (DTGS) detector or mercury cadmium telluride (MCT) detector, and the intensity is measured as a function of optical path length difference. Deuterated triglycerine sulfate detectors are room temperature detectors, which makes them a little more convenient to use, but less sensitive. Mercury cadmium telluride detectors are usually liquid nitrogen cooled, which makes them slightly more of a hassle to use, but much more sensitive.

The broadband IR light source for most FTIR spectrometers is a globar, which is a piece of silicon carbide that is heated by passing electrical current through it until it glows hot in the IR. This type of thermal emission has a characteristic ratio of intensities to wavelength and is known as "blackbody radiation" (see Appendix 4).

A background measurement is performed to determine the interferogram with no sample present. This measurement is used to correct for intensity differences at different wavelengths and also for atmospheric water vapor and carbon dioxide. FTIR spectrometer software will almost always prompt you to take a new background measurement as you start a new set of data acquisition.

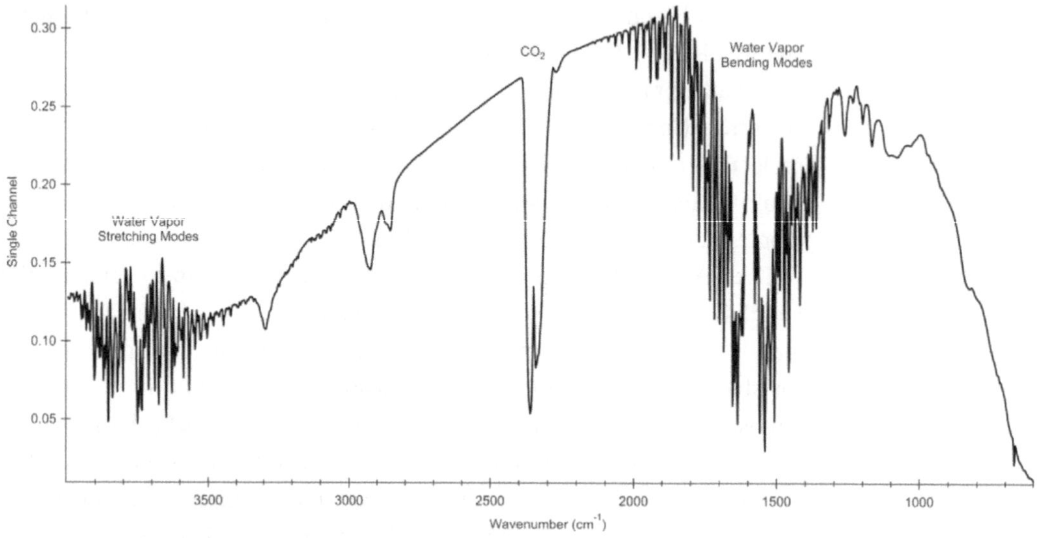

Figure 3.18

Background measurement showing the blackbody radiation shape to the intensities of the light as a function of wavelength and absorbances from atmospheric water vapor, carbon dioxide, and some coatings on the optics.

Some FTIR systems allow you to purge the sample chamber with nitrogen gas, which has no infrared absorbance as it has no change in dipole moment with its vibrational motion, or put the sample under vacuum to eliminate atmospheric water vapor and carbon dioxide with the beam path. If you are purging your sample area with nitrogen to try to get rid of the water vapor and carbon dioxide, you will need to wait the same amount of time between when you started the purge and when you took the background measurement as when you loaded the sample and started taking data so that the concentrations of water vapor and carbon dioxide in the sample area are the same. In practice, many spectroscopists just try to live with the background subtraction, compensating for the water vapor and carbon dioxide IR absorbances rather than messing around with nitrogen purges or vacuum systems.

The sample is then introduced to the beam path, and if the sample absorbs certain frequencies of light due to the molecular absorbances, then the constructive and destructive interference pattern will change. The captured interferogram is unfortunately hard for humans to interpret. The interferogram is processed with a Fourier transform, hence the FT part of FTIR.

The Fourier transform basically replots the data with frequency on the x-axis instead of optical path length difference. FTIR spectra are plotted as a frequency vs. absorbance plot (or a frequency vs. %transmittance or frequency vs. %reflectance plot) for ease of human interpretation. Frequency is traditionally plotted in units of cm^{-1} (called wavenumbers). The typical FTIR spectrum is from 4000 to

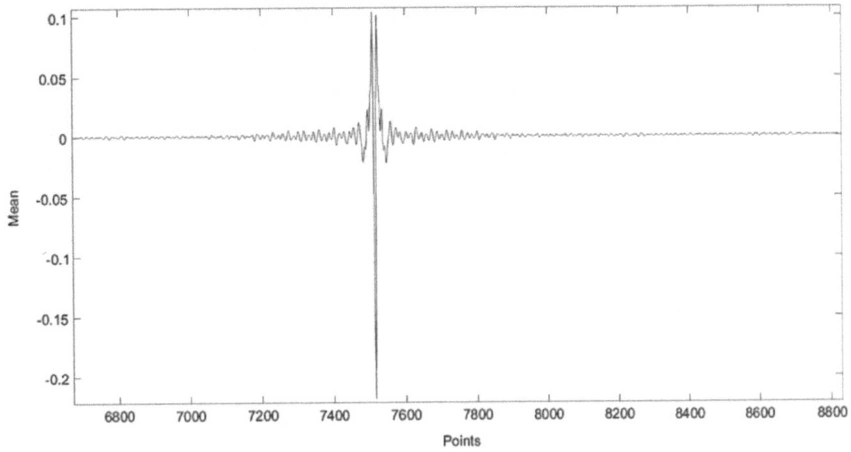

Figure 3.19

Interferogram from FTIR spectrometer.

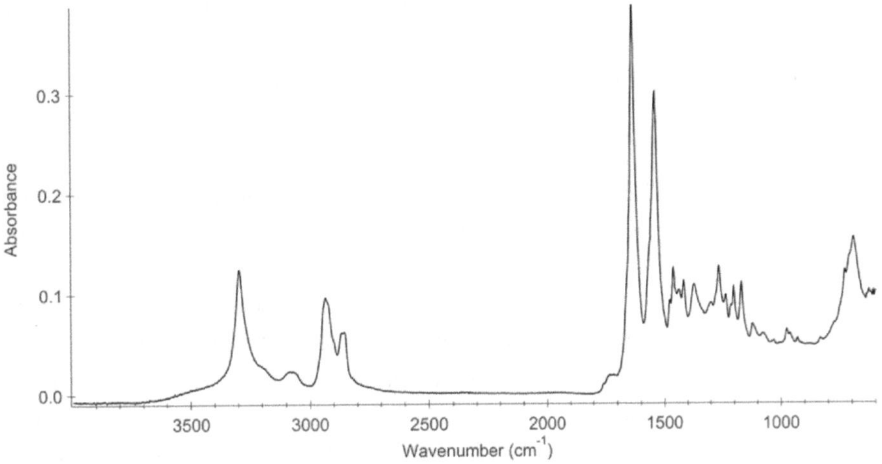

Figure 3.20

Typical FTIR spectrum. ATR-FTIR spectrum of nylon.

$400\,cm^{-1}$. Some FTIR spectrometers are starting to be able to probe into a lower frequency range down to $50\,cm^{-1}$ with new developments in beam splitter materials and automated beam splitter switching technology.

Most FTIR spectrometers default to $4\,cm^{-1}$ resolution. This is reasonable for liquid and solid phase samples, as intermolecular interactions will broaden the absorbance peaks. The only time you may want to change this is if you are doing a gas phase measurement.

FTIR can be performed in a number of different collection modes. The three most common are transmission, attenuated total reflection (ATR), and specular reflection. Most FTIR spectrometers support a range of accessories for the different sampling modes. The accessories are often made by third-party companies. Pike Technologies and Harrick Scientific Products are two of the main FTIR accessory manufacturers.

3.4.1 Transmission FTIR

Transmission was historically the first mode developed and is the easiest to think about. To perform an FTIR transmission measurement, you collect a background spectrum to take into account the instrument response function which includes the unevenness in the intensity of the light at different wavelengths, the quantum efficiency of the detector, and for FTIR the water vapor and the carbon dioxide in the air in the beam path. You then place your sample in the broadband IR light path and have the system measure which wavelengths have been absorbed by the sample. FTIR transmission spectra are commonly plotted in two different ways. FTIR spectra can be plotted either as percentage transmission vs. wavenumber or absorbance vs. wavenumber. In the percentage transmission plot, the signal comprises dips down from the 100 percent transmission line. In the absorbance plot, the signal comprises peaks coming up from the zero absorbance line. The percentage transmission plot is the real measurement. The absorbance plot is calculated and does contain the assumption that your sample is not scattering light.

To perform a transmission FTIR measurement, your sample must be thin enough not to absorb all the light at a specific frequency. The most common way is to grind a sample up and mix it with potassium bromide (KBr), which has no IR absorbance. The mix of sample and KBr is then pressed into optical windows commonly called pellets. These KBr pellets hold the powdered sample in a dilute solid solution for making the FTIR transmission measurement. However, this method tends to be a hassle and has fallen out of fashion in favor of ATR-FTIR techniques (Section 3.4.2).

TIP: KBr pellet sample holder cards are available and make handling KBr pellets a lot easier. They are basically a piece of cardboard with a 10 mm hole punched in it and double-sided adhesive. You peel off the adhesive covering, drop the 13 mm KBr pellet over the hole, and fold over the other half of the card. The cards then slide into the standard transmission sample holder of an FTIR spectrometer.

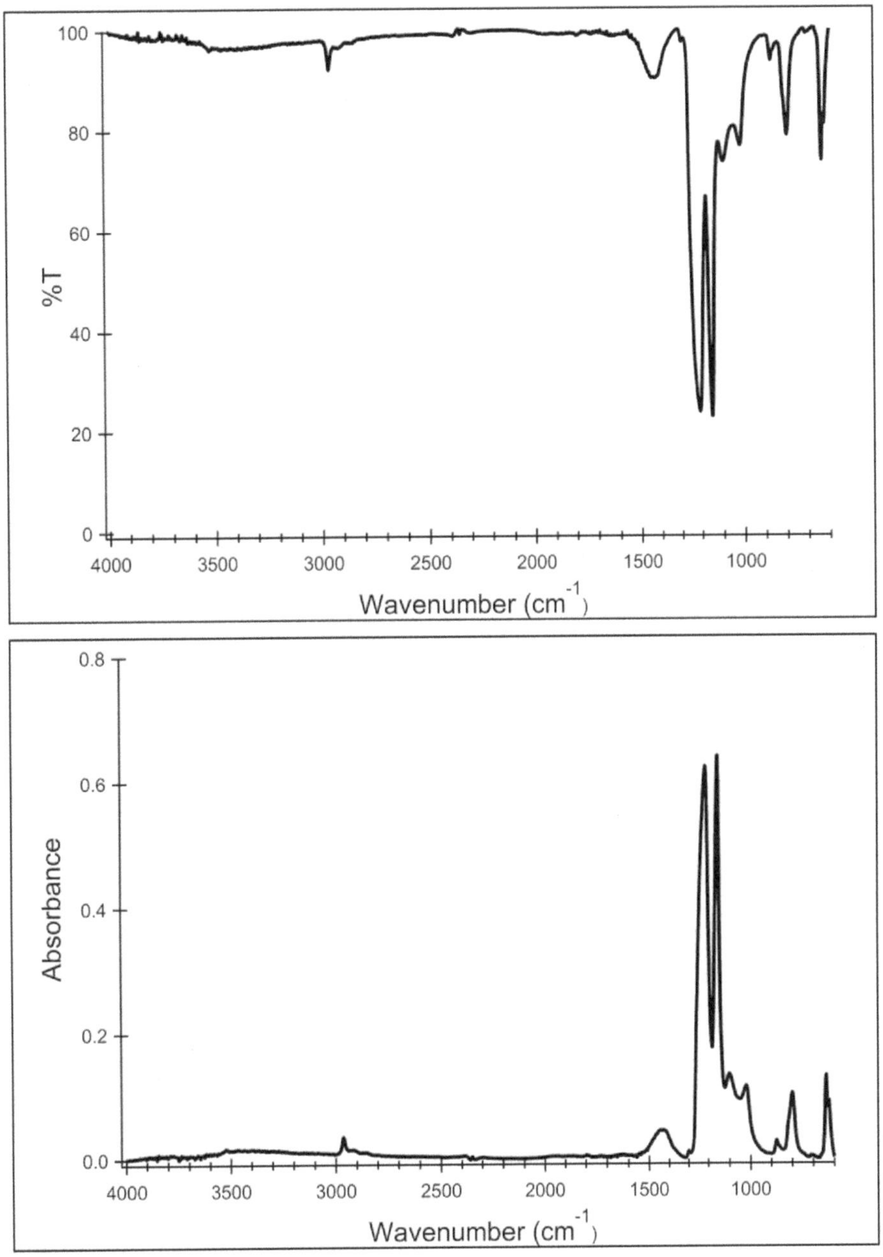

Figure 3.21

Transmission FTIR spectrum. FTIR spectra are sometimes plotted with percentage transmission (%T) on the *y*-axis and sometimes with absorbance units (Abs) on the *y*-axis.

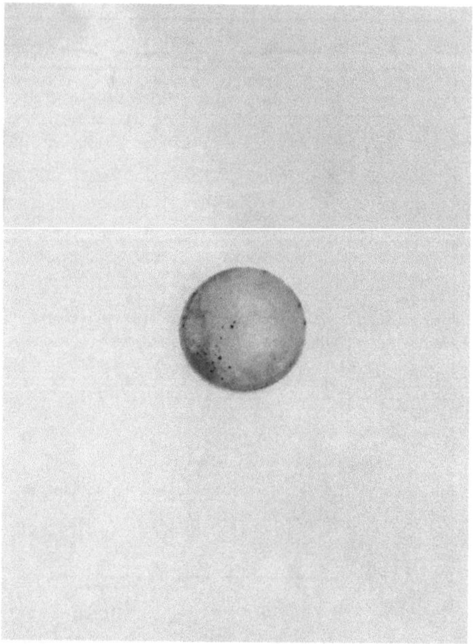

Figure 3.22

Powdered sample in a KBr window.

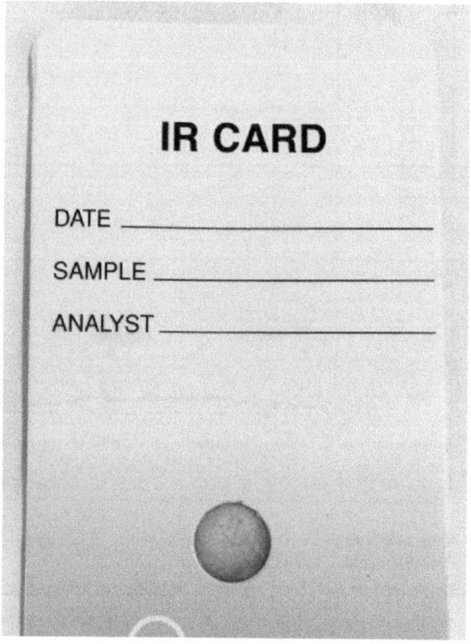

Figure 3.23

Sample holder cards make handling KBr pellets for transmission FTIR much easier.

3.4.2 Attenuated Total Reflectance Fourier Transform Infrared Spectroscopy

Attenuated total reflectance Fourier transform infrared (ATR-FTIR) is probably the most popular collection method today due to the fact that it requires essentially no sample preparation. ATR is a collection mode that uses an IR transparent crystal (Si, Ge, Diamond, and ZnSe are common) in contact with the sample. The ATR crystal is cut and oriented in such a way that the light undergoes total internal reflection at the interface between the sample and the crystal. Only an evanescent electric field from the light penetrates and interacts with the sample. The evanescent electric field probes approximately the top 500 nm of the sample, making it more sensitive to surface chemistry.

ATR crystals can be single-bounce, when the IR light interacts with the sample a single time, or multi-bounce, when the IR light interacts with the sample multiple times. A multi-bounce ATR crystal will have a lower concentration detection limit as each time the light interacts with the sample more light is absorbed, bringing the absorbance peaks up out of the noise of the spectra. A multi-bounce crystal will require a larger sample volume though.

The exact depth depends on the optical properties of the sample and the optical properties of the ATR crystal and the exact wavelength of the light. Confusingly, the attenuated total reflection spectrum is similar to a transmission FTIR spectrum and is totally different from a specular reflection spectrum. To make the ATR spectrum match the transmission spectrum of the same sample, a mathematical ATR correction needs to be performed (Section 3.4.8) on the collected spectrum to take into account the different penetration depths at different wavelengths. (Remember that we are working with light ranging from 3 to 20 μm in wavelength.) The longer the wavelength, the deeper into the sample the evanescent electric field penetrates. Thus, longer wavelengths effectively interact with a larger sample volume, causing misleadingly larger absorbance peaks at the low-frequency end of the spectrum.

You do need to be careful about cleaning the ATR crystal properly before you contact it with your sample. If any of the previous sample is sticking to the ATR crystal, you will measure the chemistry of the previous sample. The ATR crystal is an optical surface, so you should clean it like any other optical surface with lens cleaning tissue and isopropyl alcohol (Section 5.1).

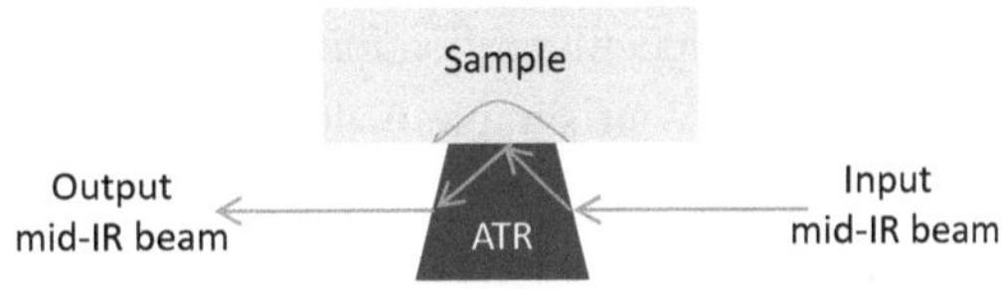

Figure 3.24

The input mid-infrared beam is sent in such that it undergoes total internal reflection at the sample/ATR crystal interface. Only an evanescent electric field penetrates the sample.

Figure 3.25

A single-bounce germanium ATR crystal with Teflon trough for liquid samples. A drop of liquid is placed on the Ge crystal and then the ATR-FTIR spectrum can be collected. The trough would also be used for powdered samples in conjunction with a pressure arm to press down on the powder and make sure there is good contact between the crystal and the powder. The trough can be removed for larger solid samples.

ATR crystals can be configured to interact with the sample from the top or from the bottom. Bottom ATR crystals typically have the option of a trough and a pressure arm, which will allow the easy measurement of FTIR spectra from liquid or powder samples.

3.4.3 Specular Reflection

Specular reflection is probably the least popular of the common sampling modes due to the fact that the reflectivity of the sample is convolved into the signal with the resonant molecular vibrational absorbance. This leads to a harder to interpret spectrum. Specular reflection does have the advantage that it is totally non-contact mode so if your sample is delicate and valuable this may be the way to go.

3.4.4 Polarized FTIR

FTIR can be performed with polarized IR light. The advantage here is that by varying the direction of the electric field of the light, some molecular orientation

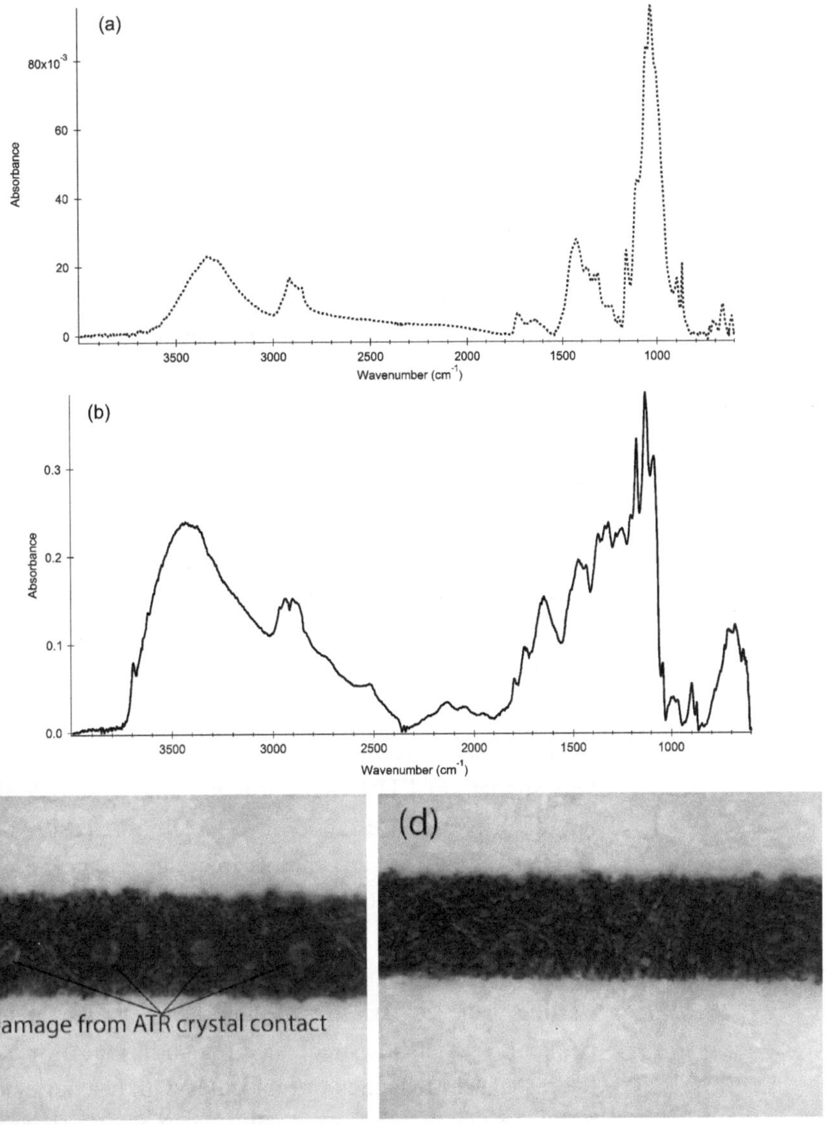

Figure 3.26

(a) ATR-FTIR spectrum; (b) specular reflection FTIR of the same sample; (c) damage of the sample due to contact with the ATR crystal; (d) sample undamaged by specular reflection FTIR. Sample courtesy of Elena Bulat.

information can be determined. When the electric field is aligned with the molecular motion, there will be a stronger absorbance of the light. Polarized FTIR measurements are often performed with a grazing angle accessory. The most common examples of polarized FTIR are determination of angles of self-assembled monolayers on gold.

3.4.5 Diffuse Reflectance Fourier Transform Infrared Spectroscopy

If your sample is very rough, diffuse reflection can be performed using an integrating sphere, which is basically a special spherical mirror that will capture all the scattered light and send it on to the detector, or with a diffuse reflectance infrared Fourier transform spectroscopy (DRIFTS) accessory. This is somewhat less common.

What Scientific Questions Can Be Asked?

Typically, FTIR spectroscopy is used to answer questions about the molecular chemistry of the sample, either to identity an unknown sample or to look for changes in the chemistry of a sample. In biology, the secondary structure of a protein can be determined through careful fitting of the amide I band. In nanotechnology, many meta-materials are designed to absorb specific frequencies in the IR spectral region.

FTIR data can be analyzed on several different levels. Some questions focus on a single absorbance peak to understand what a specific functional group of the molecule is doing. Some questions use the entire FTIR spectrum as a signature of a specific molecule. Some questions use the entire spectrum and see how a population of samples differ from each other.

Strengths and Limitations

One strength of FTIR is that it is immune to sample fluorescence. Its limitations are that the spatial resolution is typically limited to $10\,\mu m^2$ as a minimum unless you combine it with a scanning probe technique (Section 3.4.15).

Water has very strong IR absorbance bands that can present challenges for collecting good data from a hydrated sample. Historically, FTIR has been limited to $400\,cm^{-1}$ and above. The lower-frequency region (down to $\sim 50\,cm^{-1}$) is starting to be accessible with new beam splitter materials, but is still not very common yet. The low-frequency region has the advantage of being able to detect the resonance frequency of molecules with heavier elements. It has the disadvantage of running into strong water vapor rotational motion absorbance, and thus purging the beam path with nitrogen or putting the entire sample under vacuum becomes essential.

What Samples Are Appropriate?

Water has a very strong IR absorbance, so samples that are hydrated are hard to work with in FTIR. For ATR-FTIR the samples must be smooth enough or soft enough to get good optical coupling between the ATR crystal and the sample. Chemists in particular use FTIR extensively to characterize synthetic reaction products; see below and Section 3.10 for more details.

Schematics

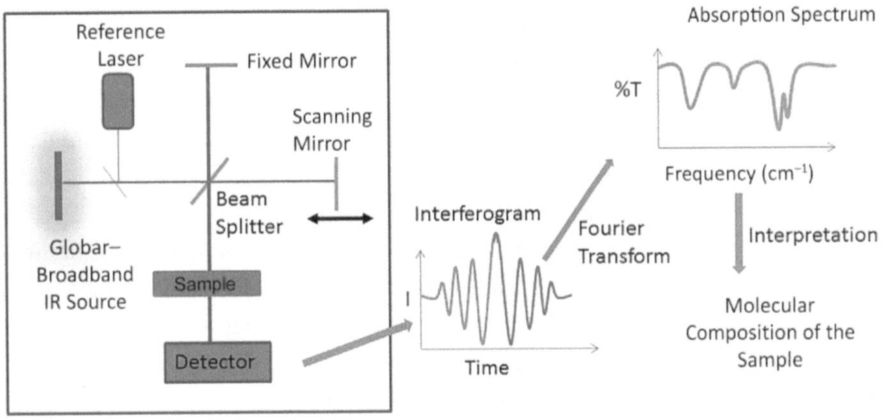

Figure 3.27

Schematic of Michelson interferometer inside a typical FTIR.

TIP: Using D_2O (deuterium oxide) instead of H_2O to prepare a sample will shift the water absorbance bands. This can sometimes make it easier to see and interpret the peaks of interest.

Common Pitfalls

Atmospheric carbon dioxide (CO_2) has two very clean well-defined peaks around 2350 cm^{-1}. If the background subtraction doesn't work perfectly, these peaks can show up in your spectra. There is often a strong temptation to interpret them as being from your sample as they look like nice clean peaks. Please don't.

Glass is not transparent in the mid-IR spectral region. Don't expect to be able to make a transmission measurement of a sample spread on a microscope slide. Some materials that are transparent in the IR are NaCl, KBr, ZnSe, double-polished Si wafers, and double-polished Ge wafers.

Most FTIR spectrometers need their desiccant to be periodically changed. If no one is taking care of a spectrometer for years and you suddenly decide to try to take a measurement with it, do not be overly surprised if you see strong water absorbances (a large number of narrow sharp peaks between 1500–1800 cm^{-1} and 3200–3600 cm^{-1}) in the data. This usually means that you need to figure out how to change the desiccant in the system. Too much water vapor for too long will actually dissolve the IR optics in your spectrometer, as they are typically made out of salts. This is why you always want to leave your spectrometer powered on. There are usually heaters in the spectrometer that help keep the optics dry.

Sample Data and Data Interpretation

Some simple data processing can help with FTIR data interpretation. As mentioned earlier, FTIR data can be plotted with wavenumbers (cm^{-1}) along the x-axis and %transmission, %reflectance, or absorbance on the y-axis. In the %T or %R plots, your signal from the molecular absorbances will be composed of dips. In the absorbance plot the signal will be peaks. The %T and %R plots are linear scales and the Abs plot is a log scale. FTIR is an absorption spectroscopy technique so, as with UV-VIS absorption spectroscopy, the Beer–Lambert law can be applied and the peaks in the Abs plot will scale linearly with the concentration of the molecular functional group (see Figure 3.21).

The actual measurement that is made in FTIR is the difference in amount of light transmitted (or reflected) relative to the background measurement with no sample. To make the absorbance plot from the percentage transmission data, the FTIR acquisition software assumes that if the light was not transmitted, then it was absorbed. This, of course, is an imperfect assumption as light can also be reflected or scattered during its interaction with the sample. This is why absolute quantification of functional groups is hard in FTIR, but if you are careful to prepare your samples in the same manner, relative quantification (e.g., Sample A has twice the concentration of C=O groups relative to sample B) is fairly straightforward. FTIR is commonly applied in synthetic chemistry for determining the products of reactions; in materials science for determining the presence of functional groups in polymers; and in artwork conservation and archeology for determining chemical compounds present in cultural heritage artifacts, etc. See most organic, analytical, or inorganic chemistry textbooks or websites such as https://en.wikipedia.org/wiki/Infrared_spectroscopy_correlation_table for data to correlate the wavenumber and shapes of FTIR peaks with specific functional groups (e.g., –OH, –COOH) in materials and molecules. We will not reproduce these extensive tables here since they are well-documented and commonly available. NIST also maintains an open-access database of IR reference spectra for countless molecules, functional groups, materials, and other samples (http://webbook.nist.gov/chemistry).

3.4.6 Baseline Correction

The light that is lost due to scattering or reflection in an FTIR transmission measurement will usually not have strong wavelength dependence and thus will appear in the data as a baseline offset or a baseline slope. This leads us to the first data-processing step of **baseline correction**. The FTIR instrument's data acquisition software will have a baseline correction function that will try to fit a curve to the baseline and then subtract off the baseline offset to produce a flat baseline with 100 percent transmission or zero absorbance, depending on the plot in spectral regions with no molecular absorbances.

3.4.7 Atmospheric Compensation

Ideally the background measurement will correct for the water vapor and carbon dioxide absorbance contributions from the atmosphere. The background subtraction is done automatically by the FTIR spectrometer software, but it is often not a perfect compensation as the water vapor and carbon dioxide concentrations in the atmosphere can vary dramatically with environmental factors as simple as if the spectrometer operator were to start talking to a colleague.

The positions of the water vapor and carbon dioxide peaks are well known and can be mathematically corrected for in post-processing. Most FTIR spectroscopy software programs have **atmospheric compensation** correction. It is often not perfect but it will help remove large CO_2 peaks that can be very distracting or even misleading to those who do not realize that they are CO_2.

3.4.8 ATR Correction

The penetration depth of the evanescent electric field has a wavelength dependence to it. The longer the wavelength, the deeper into the sample the light penetrates. The longer wavelengths interact with a larger sample volume so the absorbances measured are artificially larger. This difference between light sample volume interactions with wavelength can be mathematically corrected by an **ATR correction** algorithm:

$$d = \frac{\lambda}{2\pi\sqrt{\left(n_{crystal}^2 \sin^2(\theta) - n_{sample}^2\right)}} \tag{3.7}$$

This is usually an option built into the software on the FTIR spectrometer. You will need to tell the algorithm what the ATR crystal is made out of, what the infrared light angle of incidence is, and if you have an estimate of the index of refraction for your sample. (An estimate of $n = 1.3$ is usually reasonable for polymers and biological materials.)

It is important to report if your ATR data are corrected or not. It is better to correct all the ATR data as the penetration depth of the evanescent wave will vary with different ATR crystal materials and it would be best to be able to directly compare data that were collected on different instruments with different ATR crystals.

3.4.9 Averaging

There is already an element of signal averaging built into FTIR data acquisition as FTIR software always asks how many scans of the Michelson interferometer to make. Often you can select how long to scan in minutes rather than the number of scans if you prefer. A scan time of one minute has become the standard for the scientific literature.

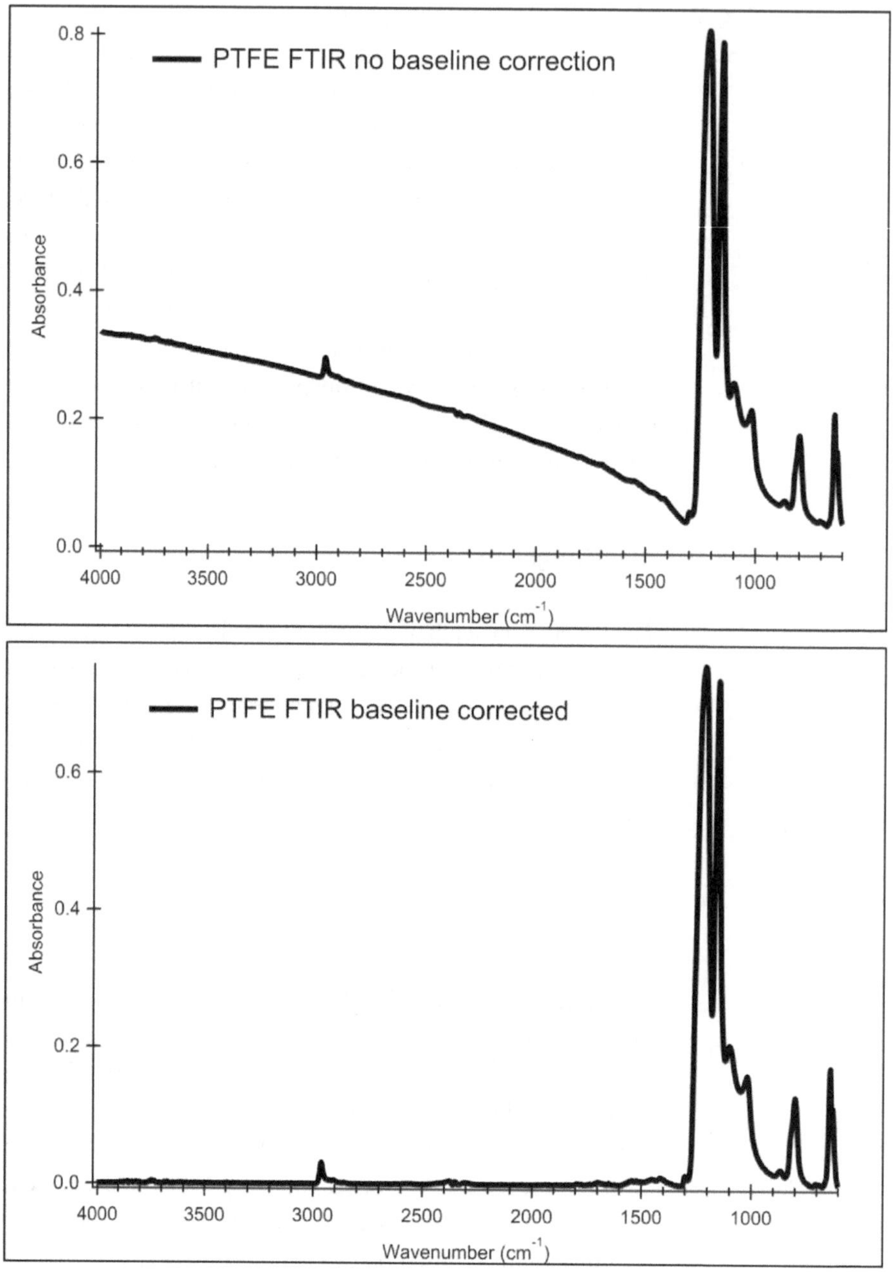

Figure 3.28

Baseline corrections for light scattered out of the light path.

One important point to understand is that the longer you scan, the smaller the noise in your signal will be. Scanning longer will never make your peak heights increase. The absorbance peak heights are determined by the molecular absorptivity and the concentration in the sample. This property of FTIR data makes it easier to compare concentrations between samples (as compared to Raman spectroscopy, where peak heights are sensitive to a number of parameters), but it also is part of the detection limit of FTIR. If you want to be sensitive to a lower concentration, you will need to have the light interact with a larger sample volume, either by having a longer transmission cell or by using a multi-bounce ATR crystal.

Averaging spectra from multiple samples together gives you a more robust spectrum as you take into account small variations in different samples.

3.4.10 Smoothing

Smoothing algorithms are usually some variation on a running average of the data in x. If you need to resort to smoothing your data you should probably just go back and collect cleaner data in the first place. While these algorithms are usually an option in the FTIR software package, you are throwing away spectral resolution for no particular benefit beyond aesthetics, and if you find that your spectra are noisy enough that you feel the need to smooth them, then you probably need to take better data if at all possible.

3.4.11 Difference Spectra

The human eye is naturally drawn to the largest feature in the spectrum. Often subtle yet important differences between spectra can be missed by casual human inspection. One way to highlight subtle differences between two spectra is to normalize the spectra (set a peak height that should be the same in each sample to 1) and subtract one spectrum from the other. This result is usually referred to as a difference spectrum. Some parts are positive and some parts are negative, depending on exactly how the spectra differ. The parts of the two spectra that are the same are canceled out with the normalization and subtraction. If there is a peak position shift the result in the difference spectrum will be a derivative shape. What you really care about in this case is how these two spectra are different, not how they are similar.

A variation on the difference spectrum for a population of spectra is the mean centered spectrum, where you average all the spectra together and then subtract the average from each individual spectrum. Again this method highlights how the spectra are different from each other.

3.4.12 Chemometrics and Multivariate Image Analysis

Chemometric and multivariate image analysis techniques are immensely powerful ways to deal with large numbers of spectra. As a human, you tend to focus on the

Figure 3.29

FTIR map showing the intensity of a C=O bond stretch overlaid on a business card.

largest peaks in the spectrum. Computers are not distracted as easily. Since chemometric techniques are widely applicable, they are discussed further in Section 3.9.

3.4.13 FTIR Mapping

Chemical mapping of a sample can be performed with FTIR imaging. The most common way is point-by-point mapping. That is, you take a spectrum from a small point on the sample, move the sample a small known amount with a motorized stage, and then take another measurement. This is a slow process. Some systems do have focal plane arrays to collect spectra from multiple positions at once, but focal plane arrays tend to be very expensive. FTIR chemical imaging is an optical diffraction-limited technique, which means that the smallest area that a spectrum can be collected from is ~10 μm square.

3.4.14 Quantum Cascade Lasers as FTIR Light Sources

Quantum cascade lasers (QCLs) are a new and much brighter source that is starting to enter the FTIR imaging and spectroscopy world. The bright source allows for shorter acquisition times at each point. The downside is that the QCLs currently available have a limited spectral range to ~100 cm^{-1}. Remember that a full FTIR spectrum is 4000 to 400 cm^{-1}. This means that QCLs typically are only good for applications where a specific known peak is being characterized. Some laser systems combine up to four QCL chips in a single package, but that is still a very limited tuning range compared to a broadband globar source.

3.4.15 Coupling Scanning Probe Techniques with FTIR Spectroscopy

To beat the diffraction limit of light (~10 μm^2 spot size for mid-IR spectral region) but maintain chemical mapping capabilities, IR lasers have been combined with scanning probe techniques to perform scattering scanning near-field optical microscopy (S-SNOM) and photo-thermal expansion microscopy. The spatial resolution of these techniques depends on the sharpness of the metal-coated AFM cantilever being used, which is usually in the 10–20 nm range.

In S-SNOM, an IR laser (usually a QCL) is focused down onto a metal-coated AFM tip. The AFM tip then taps on the sample. When it is close to the sample there is a near-field enhancement of the electric field leading to a larger absorbance of IR light by the sample. Both the far-field and near-field signals are collected, then sent through a Michelson interferometer and a lock-in amplifier locked to the tapping frequency of the tip. The lock-in amplifier collection technique is used to separate the near-field and far-field contributions to the signal.

In photo-thermal expansion spectroscopy, an AFM tip is brought into contact with the sample and an IR pulse of light is shone on the sample. If the sample absorbs that wavelength of light, it will heat due to the absorbed energy from the laser and expand. The sudden expansion of the sample will cause the AFM tip to oscillate. The IR laser is tuned across the infrared range and the amplitude of AFM tip oscillations can be plotted as roughly the equivalent of an FTIR spectrum from the area that the AFM tip was in contact with the sample.

The biggest limitation to these techniques currently is the speed and range over which the IR lasers can be tuned. The slowness of the laser tuning makes it impractical to collect a full spectrum at each point. Typically either a single point is selected and a full spectrum is collected from that point or a single wavelength of the IR laser is selected and the sample's response to that particular wavelength is mapped. Currently, 900 cm^{-1} is the lowest frequency that either mid-IR OPOs or QCLs can reach.

3.4.16 Infrared Tomography

Since IR spectroscopy is an absorption phenomenon, tomography techniques can be used to generate a three-dimensional chemical map of the sample.

Tomography techniques involve taking a series of absorption images from every angle of a sample and then back-calculating the internal three-dimensional structures. Tomography is commonly done with x-rays and is referred to as micro-computed tomography (μCT). (In a medical setting, x-ray tomography is also sometimes called a computerized axial tomography [CAT] scan.) Tomography techniques can also be applied to transmission electron microscopy. Infrared tomography requires a very bright source of IR light and is currently only being implemented at certain synchrotron sources such as the Berkeley lab's Advanced Light Source.

3.4.17 Sum Frequency Generation Vibrational Spectroscopy

If only the surface chemistry of the sample is of interest then sum frequency generation vibrational spectroscopy (SFG-VS) is a useful technique. SFG is a second-order nonlinear optical technique that uses a tunable pulsed mid-IR laser and typically a fixed wavelength pulsed visible laser. When the two laser pulses arrive at the same time, a photon with the sum of the two input frequencies can be generated. The probability that the sum frequency photon is generated is much higher when the IR laser is tuned to a vibrational resonance of the sample. SFG-VS is a simultaneous IR absorption and Raman scattering phenomenon. To be SFG VS active, the normal mode of the molecule must be IR and Raman active.

To perform narrowband SFG-VS experiments, usually a fixed wavelength visible pulsed laser beam is overlapped in time and space on the sample with the tunable IR pulsed laser beam and the intensity of the sum frequency generation signal is measured with a spectrometer and a PMT. The IR laser is then tuned to 5 cm^{-1}, the spectrometer is tuned, and the intensity of the SFG-VS signal is measured again. One benefit to the SFG-VS technique is the signal is shifted up in wavelength so the signal is immune to interference from fluorescence.

The biggest advantage of SFG-VS is that being a second-order nonlinear process (a $\chi^{(2)}$ process in physics jargon), the sum frequency photon can only be generated from a non-centrosymmetric location in the sample. A non-centrosymmetric location is a place where up and down look different to a molecule. Surfaces and interfaces are by definition non-centrosymmetric locations in a sample. Surfaces are where a sample interacts with the rest of the universe. Surface chemistry has long been an area of study, but most of the other techniques developed require ultra-high vacuum (UHV) systems. Soft samples such as polymers or biological samples can reorient and change when placed in a UHV chamber. Being an optical technique, SFG-VS doesn't require a UHV chamber and provides molecular as opposed to atomic composition information. SFG-VS can be used to probe buried interfaces too, as long as photons can reach the interface.

Another advantage of SFG-VS is that it is a polarized light technique, which allows you to determine the orientation of the functional group on a surface by measuring the response with a couple different polarization combinations of excitation and signal beams. Whether a functional group is standing up or laying down on a surface can have dramatic implications for the surface interactions of a sample.

SFG systems can be homebuilt or purchased from a number of vendors.

3.5 Raman Spectroscopy

Spontaneous Raman spectroscopy is the other commonly used vibrational spectroscopy technique. Spontaneous Raman spectroscopy systems can be homebuilt or purchased as a full system. Raman has the convenient property of working with visible light as an excitation source, and offers complementary information to FTIR since the selection rule for Raman activity is for the molecular vibrational mode to have a change in polarizability and the selection rule for FTIR activity is for the molecular vibrational mode to have a change in dipole moment.

Underlying Physical Principles

Raman is a scattering phenomenon, as opposed to FTIR which is an absorption phenomenon. When a photon interacts with a sample, it will occasionally lose a small quantum of energy to a vibrational mode of the molecule. (Please refer to the discussion of where quantized vibrational energy levels come from in Section 3.4.) This loss of energy will cause the photon to shift in wavelength (change color) before it is scattered from the sample. The identity of the vibrational mode that the photon interacted with is encoded in the difference in energy ("shift") between the incident and scattered photons.

Raman scattering is a very weak phenomenon. Approximately one photon in 10^8 excitation photons will undergo Raman scattering. A number of techniques attempt to enhance the signal, such as resonance Raman (Section 3.5.12), CARS (Section 3.5.13), SRS (Section 3.5.14), SERS (Section 3.5.15), and TERS (Section 3.5.16).

Since the information about the chemical composition of the sample is encoded in the shift in energy of the scattered photon from the energy of the excitation photon, it is important to start with all of the excitation photons having the exact same energy for optimal spectral resolution. For this reason, narrow line continuous wave gas lasers or diode pumped solid state lasers are preferred excitation sources for Raman spectroscopy. Note that diode laser emission can drift in wavelength by several nanometers if the temperature of the laser head is not precisely controlled.

Since the information we are interested in is a shift from the excitation (rather than the magnitude of the photon energies themselves), we are free to choose

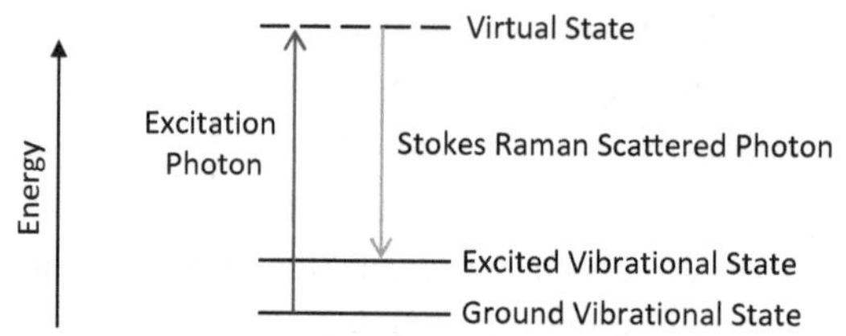

Figure 3.30

Energy level diagram of spontaneous Raman scattering. An excitation photon excites the molecule up to a virtual energy level. The molecule cannot stay in a virtual energy level and so immediately relaxes. The molecule occasionally relaxes to an excited vibrational energy, producing a photon of a different energy than the excitation photon. By measuring the difference in energy between the excitation photon and the emitted photon, one can determine the difference in energy levels.

different excitation lasers and we should get the same information. (Resonance Raman spectroscopy [Section 3.5.12] is an exception to this concept.)

The selection rule that determines whether a molecule is active in Raman spectroscopy is that there must be change in the polarizability of the molecule with the vibrational motion. The polarizability can be thought of as the sloshing of the electron cloud surrounding the molecule. The derivation of the selection rule is beyond the scope of this book, but we refer readers to Albert Cotton's *Group Theory* textbook for more thorough derivations and discussions.

Raman scattered photons are sorted by color with a diffraction grating and sent to a CCD camera. This is known as dispersive Raman spectroscopy. (Fourier transform [FT] Raman can be performed, but it is typically only used with a near IR laser as an excitation source and is less common.)

A number of parameters that you typically need to set on a Raman spectrometer before making a measurement are discussed in the following sections.

3.5.1 Excitation Laser Selection

The first parameter to select is which excitation laser to use for your experiment. Raman spectroscopy systems often have two or three excitation lines available to allow you to move the Raman spectra away from any florescence emission in the sample. With an unknown sample, you will typically have to pick one excitation line and see what the signal looks like. Do you get a nice clean Raman signal (narrow, sharp peaks, typically 5–10 cm^{-1} wide), or do you get a lot of fluorescence (broad peaks hundreds of cm^{-1} wide)? If the latter, try a different excitation line.

Raman scattering probability scales as ω^4, where ω is the frequency of the excitation light; so, by moving to shorter wavelengths you will collect a stronger signal. However, the probability of exciting fluorescence in the sample usually

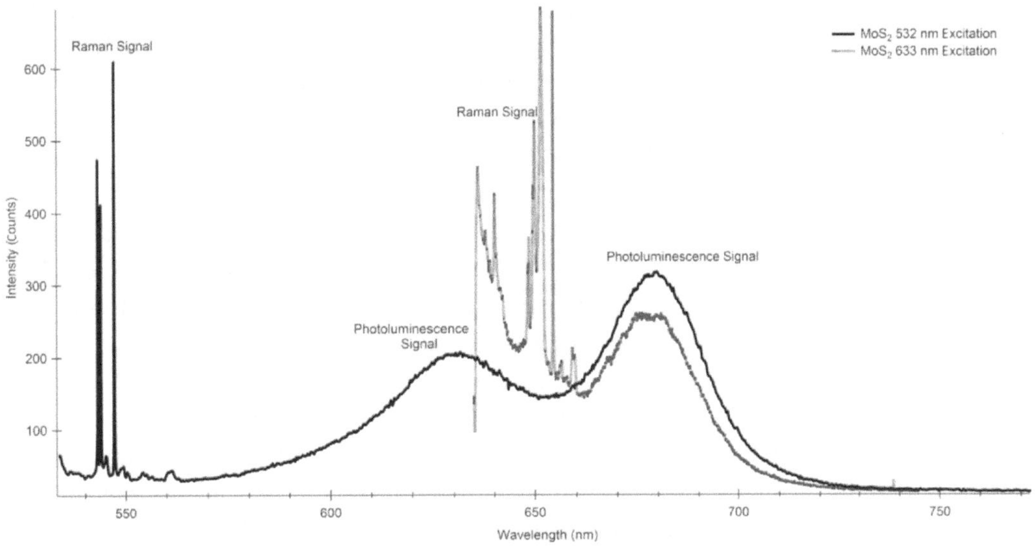

Figure 3.31

Raman and PL signals of MoS_2 sample excited with 532 and 633 nm light. Raman peaks are shifted from the excitation line. PL peaks are always in the same spectral position and tend to be much wider. Sample courtesy of Luis Jauregui.

increases as you move to shorter wavelengths. It is possible to move to short enough wavelengths that you can squeeze the entire Raman spectrum in before the start of the fluorescence emission. In Figure 3.31 you can see that by shifting to a shorter wavelength of 532 nm, the narrow Raman lines are unobstructed by the broad PL peaks at 633 and 680 nm.

3.5.2 Diffraction Grating Selection

The next choice is usually which diffraction grating to use. Raman spectrometers are often equipped with several diffraction gratings that can be switched in and out of position. (See Section 2.3.36 for a review of diffraction gratings.) The higher the blaze (grooves/mm) of the grating, the larger the angle the Raman scattered light will be spread spatially. The digital camera that is typically used as a detector in a modern Raman system has a fixed width to each pixel. So, each pixel will collect light from a narrower section of the rainbow generated by the diffraction grating, giving you better spectral resolution. Higher resolution is great, but it comes with a cost: Raman scattering is a weak signal and by spreading the light out over more pixels, each pixel will collect fewer photons. This means that the signal intensity will drop significantly with a higher blazed grating. You can compensate for the lower intensity on each pixel by increasing the exposure time (Section 3.5.3), but then your experiment will take longer.

To put it in concrete terms, switching from a 600 gr/mm to an 1800 gr/mm diffraction grating will provide approximately three times better spectral resolution but peak heights will drop by a factor of three for the same exposure time. You can compensate by increasing the exposure time, but that naturally increases the time it takes to collect data. Depending on the information that you are interested in, you may not need the higher spectral resolution.

If you are looking for stress or strain in a material then you will be looking for a 5–10 cm^{-1} shift in peak position and you will need the higher spectral resolution. However, if you are trying to identify an unknown material, you will probably not need the higher spectral resolution and are better off choosing the lower blazed grating.

The higher blazed grating will also increase the number of spectral windows that you need to collect to build up a full Raman spectrum since the camera chip has a fixed finite width. Again, using the 600 gr/mm and 1800 gr/mm example, to build up a full spectrum you will need to collect four spectral windows with the 1800 gr/mm grating to have the same spectral range as with the 600 gr/mm grating. So all together to collect the same spectral range with the same intensity peaks your experiment would take 12 times longer. To save time, spectroscopists often use the wide range grating to collect the first spectrum from an unknown sample to identify the energy range in which peaks of interest are located. Then, experimenters shift to the higher blaze grating, and only collect fewer but higher-resolution spectra in that narrower spectral window of interest. The accuracy of the measurement of the position of the Raman peaks depends on the accuracy of the positioning of the diffraction grating. Calibration of the grating position in Raman spectrometers is often performed by measuring the Raman spectrum from a silicon or germanium wafer. Bulk Ge and Si exhibit sharp, strong peaks corresponding to Ge–Ge or Si–Si bond stretching at 300 and 520 cm^{-1}, respectively.

3.5.3 Exposure Time

Unlike FTIR spectroscopy, in Raman spectroscopy experiments, you can choose how long to collect photons with the camera. There is a direct linear relation between the peak heights and the exposure time. If you double the acquisition time, you double your peak heights.

The maximum exposure time is limited by the bit depth of your camera. (See Section 2.3.47 for a review of cameras.) For a 16-bit camera, you cannot collect more than 65,000 counts on any single pixel. Raman scattering is so weak that trying to use the full dynamic range of the camera for each spectrum collected is usually impractical. You do want to be sure to have a good signal-to-noise ratio in your data. Increasing your exposure time will increase your signal. Increasing the number of measurements that you average together (Section 3.5.4) will decrease your noise. In most cases, when setting up a Raman experiment you are better off increasing your exposure time rather than averaging more measurements together.

3.5.4 Number of Measurements Averaged Together

The noise in Raman measurement is largely from noise on the camera. Silicon-based camera technology is well developed and has a relatively low readout and dark noise, so averaging a large number of measurements together usually does not provide much of an advantage toward increasing signal-to-noise ratios in your measurements. That said, it is generally worthwhile setting the number of measurements to two. If you make two measurements with the exact same conditions then you should obtain the same data each time, apart from statistical shot noise.

Cosmic rays yield one exception to the above that is particularly relevant with Raman spectroscopy. High-energy photons from outer space (x-rays and gamma rays from binary stars, etc.), collectively called **cosmic rays**, will periodically interact with a single pixel on the camera for one of the exposures. The cosmic ray will saturate one pixel for one exposure. The cosmic rays will be very strong, dramatically out of the Raman signal level. They will also be very narrow, typically only one or two pixels wide. By recording two sequential measurements, it is very easy for a computer algorithm to remove the cosmic rays from your data for you. Most Raman data-collection software packages have cosmic ray removal options built in.

3.5.5 Binning

Binning refers to adding together multiple pixels to increase the counts per pixel. The vertical pixels on the camera of the spectrometer are typically binned already. The horizontal pixels can also be binned to increase the peak intensity at the cost of spectral resolution. In some cases, the Raman signal is so weak that it is beneficial to sacrifice some spectral resolution for increased signal intensity.

What Scientific Questions Can Be Asked?

Raman shifts are characteristic of the molecular composition and chemical bonding of a sample, so Raman is widely used for identification of an unknown sample by matching with Raman spectral libraries. There is no table that will tell you what all the peaks in your spectrum are though. Atomic composition analytical techniques such as x-ray photoelectron spectroscopy (XPS; see Appendix 1) or energy-dispersive spectroscopy (EDS; see Section 4.19) will neatly label each peak with an elemental name. Atomic composition techniques only have ~120 elements as options. Raman libraries are unable to neatly label all the peaks with identities as there are millions of molecular compounds, each with multiple peaks. This means that Raman data comprise a very rich data set, but are also harder to interpret.

For instance, Raman spectra contain information about the crystal structure of the material and strain in the lattice of the material. Polarized Raman spectroscopy can be used to deduce the orientation of molecules in the sample. Raman is often used as a screening tool to determine whether carbon nanotubes are semiconducting or metallic, and is often applied to measure the quality of the lattice of two-dimensional materials such as graphene, MoS_2, BN, WSe, other transition metal (di)chalcogenides, etc.

Strengths and Limitations

One of the biggest limitations of spontaneous Raman scattering techniques is that the Raman signal can sometimes be completely obscured by fluorescence or photoluminescence from the sample. Metals also have no Raman signal, which is sometimes a limitation of the technique but can sometimes be exploited to isolate one part of the sample from another.

Recently, optical filters used for Raman spectroscopy have become very good; a standard edge filter can provide a $50\,cm^{-1}$ cutoff. Recently developed Volume Bragg Grating filters can even give spectral cutoffs as low as $5\,cm^{-1}$, meaning that even tiny shifts in energy from the excitation beam can be measured! This lower-frequency region opens up the type of samples to include molecules with heavier atoms, such as TiO_2, as compared to the current standard minimum cutoff of $400\,cm^{-1}$ in FTIR spectroscopy.

What Samples Are Appropriate?

Raman is widely applicable to many materials that have a change in polarizability with molecular motion. Of note, metals don't have any Raman signal. Metal oxides typically do have a Raman signal.

Any fluorescence or photoluminescence from the sample will make it hard to see and interpret the Raman signal. Chemically complex samples (such as biological samples) can be quite challenging to interpret with Raman because they produce many peaks. Chemometric approaches (Section 3.9) are probably the best way to approach such complicated data sets.

Common Pitfalls

Fluorescence and photoluminescence are common issues with samples. Shifting to a different excitation wavelength can help separate the Raman signal and the fluorescence signal (Figures 3.31 and 3.34).

Accidentally burning or modifying your sample with too much excitation laser power is another common issue. If your first spectrum and your second spectrum from the same spot look different, you probably modified your sample with too much excitation laser power. Typical laser power levels for micro-Raman are in the 1 mW range. (This is about as powerful as a standard laser pointer.)

Know what the spectrum of the room lights looks like. Fluorescent lighting has sharp emission lines that can be mistaken for Raman lines. There is a line from the fluorescent lights at 612 nm that often shows up when using a 532 nm excitation for Raman spectroscopy. There are also a number of lines in the red from the standard fluorescent light bulb that show up when using a 633 or 785 nm excitation laser. So always be sure to record background spectra on blank substrates and use those as a baseline before collecting data on your sample.

There are many parameters to set in a Raman spectroscopy system. Understanding all of those settings and tuning them appropriately can yield optimal results.

Schematics

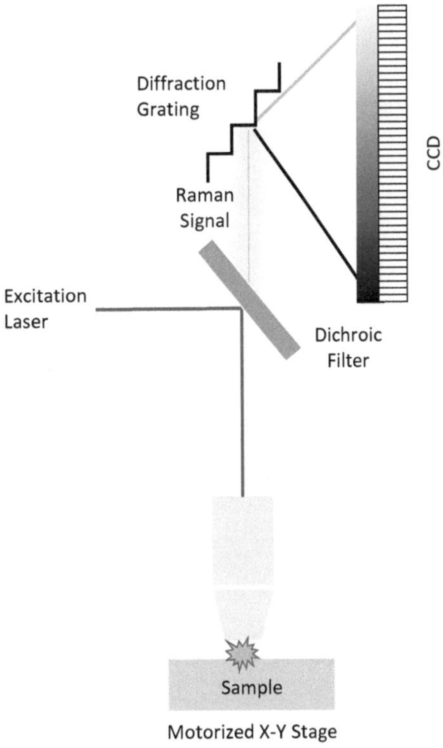

Figure 3.32

Schematic of a micro-Raman spectroscopy system.

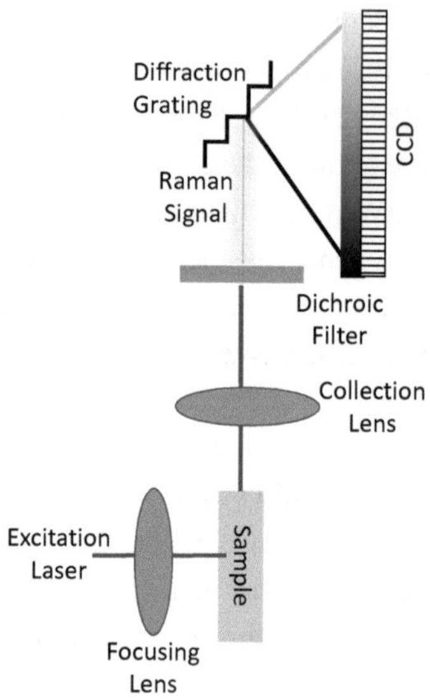

Figure 3.33

Schematic of macro-Raman setup. Collection at 90 degrees helps eliminate excitation light.

Sample Data and Data Interpretation

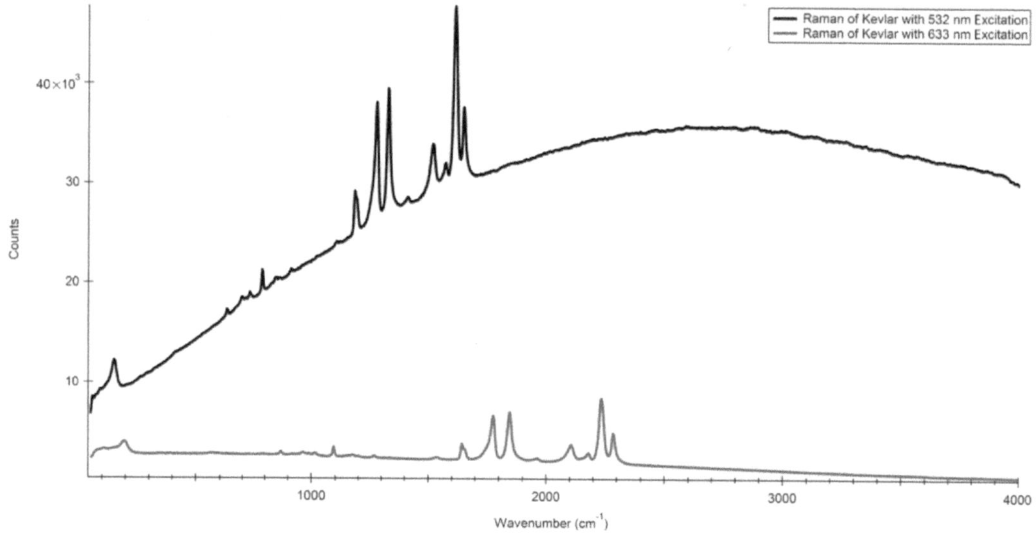

Figure 3.34

Raman spectra of Kevlar at 532 nm excitation and 633 nm excitation. The 532 nm excitation causes fluorescence in the sample so switching to the 633 nm excitation can give a cleaner signal.

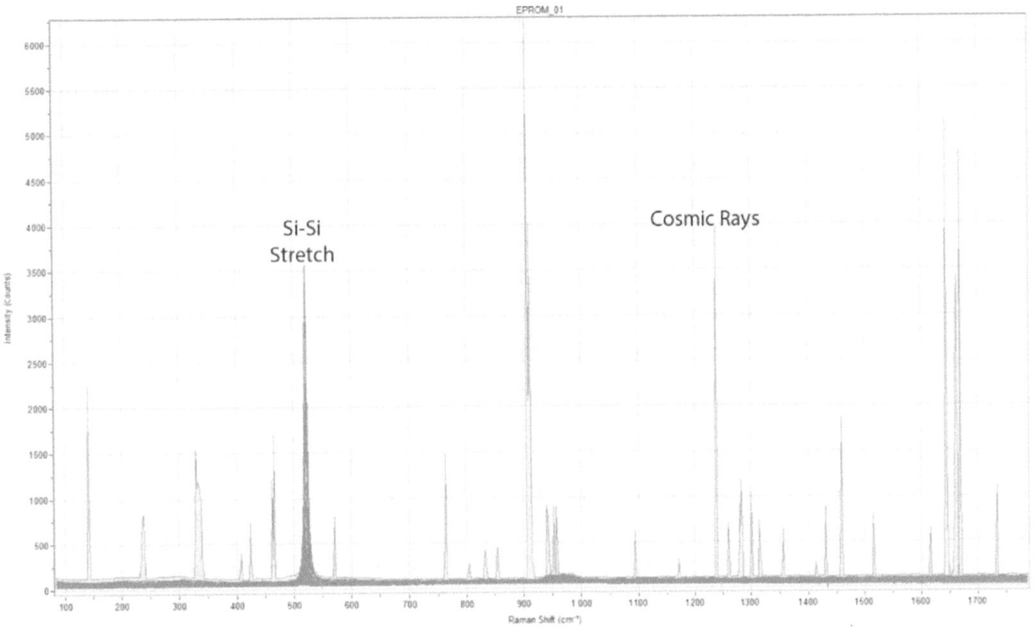

Figure 3.35

Cosmic ray spikes in Raman data.

3.5.6 Cosmic Ray Removal

In your data processing, be sure to remove any cosmic rays. These are sharp, narrow spikes in your data from high-energy photons from outer space such as gamma rays. The easiest way to remove them is to have the software compare two sequential measurements and look for strong narrow spikes that only appear in one of the two measurements.

3.5.7 Baseline Correction

Photons from fluorescence/photoluminescence emission, the room background (e.g., fluorescent lights per above), and dark counts on the detector all contribute to a baseline offset in Raman data. It is common practice to fit the baseline either with a polynomial function or a series of line segments and subtract off this baseline offset.

3.5.8 Peak Fitting

Peak fitting enables you to quantitatively report your Raman measurements. It is common to fit the peaks of a Raman spectrum to Lorentzian functions. This provides the peak center, peak amplitude, and peak width. The peak center and the peak width are more robust measurements than the peak amplitude. Raman signals' amplitudes are sensitive to many of the experimental parameters such as laser power, exposure times, how well the excitation laser is focused, the concentration or thickness of the analyte in question in the sample, scattering from surface roughness or anisotropy in sample structures, Raman scattering cross-sections of the mode in question, etc.

Peak fitting allows you to deconvolve the contributions from overlapping peaks. The sum of the individual peak fits should match your raw data. It is common to apply a baseline correction to subtract off any fluorescence contribution before attempting to fit Raman data.

3.5.9 Polarized Raman Spectroscopy

The direction of the electric field of the excitation light relative to the molecular motion of the sample can change the intensity of the Raman scattering significantly. This effect can be used to deduce the crystal orientation of the sample.

TIP: Lasers are typically linearly polarized. This means that if your sample is crystalline, you will be performing a polarized Raman experiment unintentionally. The orientation of the crystalline sample as you put it in the Raman microscope or spectrometer will have an effect on the intensity of the Raman signal. You can easily do a quick check of whether your excitation laser is linearly polarized by placing a polarizer into the beam path (a cheap pair of polarized sunglasses will do in a pinch) and rotating the polarizer. If the intensity of the beam varies after the polarizer as you rotate the polarizer, the excitation laser is linearly polarized. You may want to consider inserting a polarization scrambler into the excitation beam path to get unpolarized excitation light or a quarter-wave plate to get circularly polarized excitation light so that the orientation of the sample does not affect the intensity of the Raman signal. If you have samples that are anisotropic in any manner, the orientation of your sample on the sample stage can have a dramatic effect on the intensity of the Raman signals.

The Raman scattered light can also have a strong polarization. Measuring the depolarization ratio, which is the degree to which the polarization of the incident

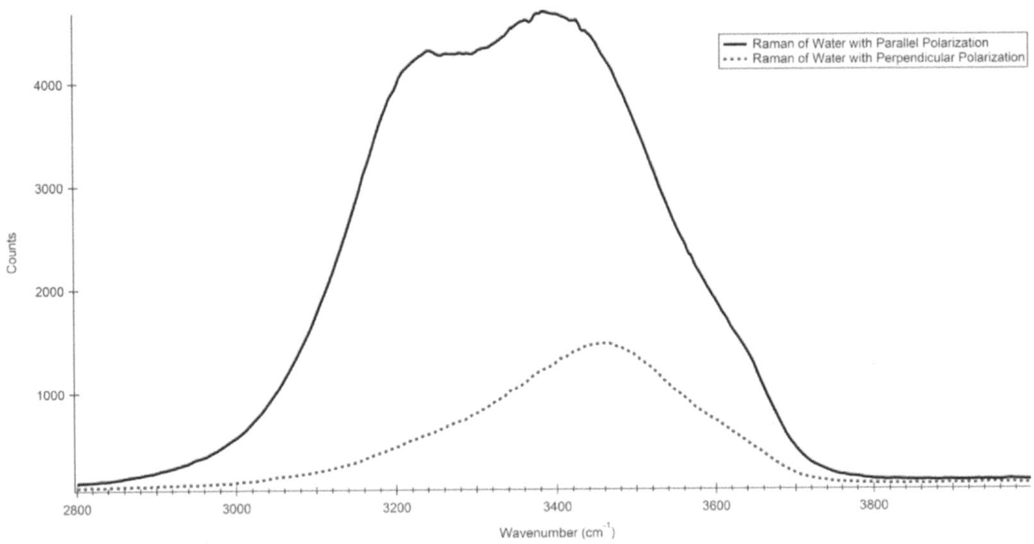

Figure 3.36

Polarized Raman measurements of water.

light is preserved in the scattered light, can provide information about the symmetry of the mode generating a particular Raman peak. This type of measurement is usually done with two polarization combinations: the excitation and the Raman scattered light with parallel polarizations and the excitation and the Raman scattered light in perpendicular polarizations.

If the polarization of the Raman scattered beam is the same as that of the incident beam (i.e., the excitation and Raman scattered light polarizations are parallel), then the Raman line is said to be *polarized.* If the Raman scattered light is intense in both the parallel and perpendicular directions, then the Raman line is *depolarized.*

Modes that are totally symmetric vibrations (a normal mode with all characters = 1 for those familiar with group theory) give rise to polarized Raman lines.

The depolarization ratio is defined as:

$$\rho = \frac{I_\perp}{I_\parallel} \tag{3.8}$$

For quantum mechanical reasons beyond the scope of this book, a mode is considered depolarized if $\rho \geq 0.75$. A mode is considered polarized if $\rho < 0.75$.

Going back to the example of the modes of water, the symmetric stretch will be strongly polarized and the antisymmetric stretch will be depolarized. So, by making a parallel polarized Raman measurement and a perpendicular polarized Raman measurement, it becomes obvious which peak arises due to the symmetric stretch.

TIP: Remember to always have a polarization scrambler in front of your spectrometer when you are trying to make polarization-dependent measurements, as the diffraction grating in the spectrometer has a much higher efficiency for light perpendicular to the grating as compared to light that is parallel to the grating. The polarization scrambler makes the light that you send into the spectrometer unpolarized so that you measure signal intensity differences due to the changes in polarization of light after interaction with your sample and not only the difference in throughput for the different polarizations of the diffraction grating. The differences in efficiency between the two polarizations depends on the grating, but can be as high as six times difference.

TIP: When doing any kind of polarization measurement, remember that any sort of plastic container tends to have some birefringence to it (Figure 1.17). This birefringence will scramble the polarization of the light, so your polarization measurements become meaningless.

3.5.10 Raman Mapping

Raman imaging can provide chemical maps of a sample similar to FTIR imaging. This is usually done in a point-by-point mapping mode, which can make it slow. You can move the laser beam with a set of scanning mirrors and keep the sample position fixed, or you can keep the laser position fixed and move the sample. One thing to keep in mind when setting up a Raman map is that you collect Raman signal from where the laser is focused (a diffraction-limited ~300 nm spot). If you set up your map to collect a Raman signal every two microns you are skipping over lots of your sample and risk accidentally missing a rare event in your sample. On the other hand, recording a spectrum at 300 nm pixel sizes takes prohibitively long, so combining the laser scanning motion and sample motion is sometimes a good approach. For instance, if you have the laser raster over a 2 μm area for each exposure and then move the sample 2 μm for the next exposure, you would be sure to have full coverage of your sample and would not miss any rare event in your sample. After the data are acquired, a peak of interest is fit for every pixel and the fit of that peak is displayed on a point-by-point basis to form the chemical image.

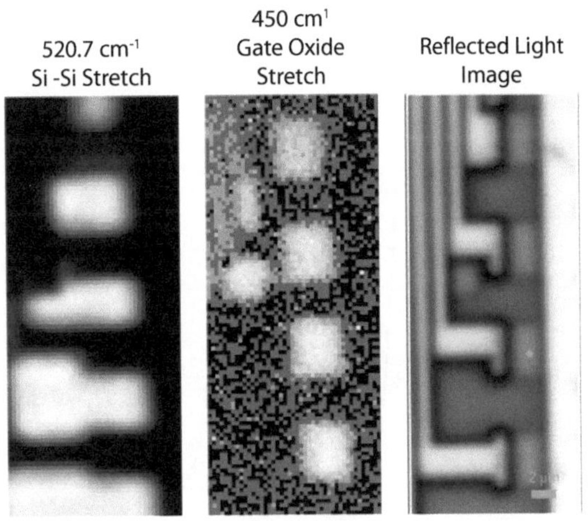

Figure 3.37

Raman map of electronic circuit showing differences in the silicon and oxide layers. The metal bus lines do not generate any Raman signal.

3.5.11 Confocal Raman 3D Imaging

Three-dimensional chemical maps can be obtained by confocal Raman imaging. The details of confocal microscopy in general are left to the discussion in the next chapter on confocal microscopy (Section 4.9). The general idea, though, is that to restrict in z where you collect your Raman signal from, map in x and y, translate in z, and then repeat until you have a three-dimensional chemical map of your sample. There is the clear limitation that you have to be able to get visible light into and out of your sample for this process to work, so it has to be at least somewhat transparent to photons around the excitation wavelength. The depth into the sample is limited by the optical scattering properties of your sample and the working distance of your microscope objective.

3.5.12 Resonance Raman

As discussed above, spontaneous Raman scattering is an incredibly unlikely event to occur with just one photon in 10^8 undergoing Raman scattering. However, there are some ways to increase the probability of Raman scattering occurring. One is referred to as resonance Raman. When the excitation light is in resonance with an electronic transition of the sample, there will be an increased probability of Raman scattering occurring. The challenge with resonance Raman is that the probability

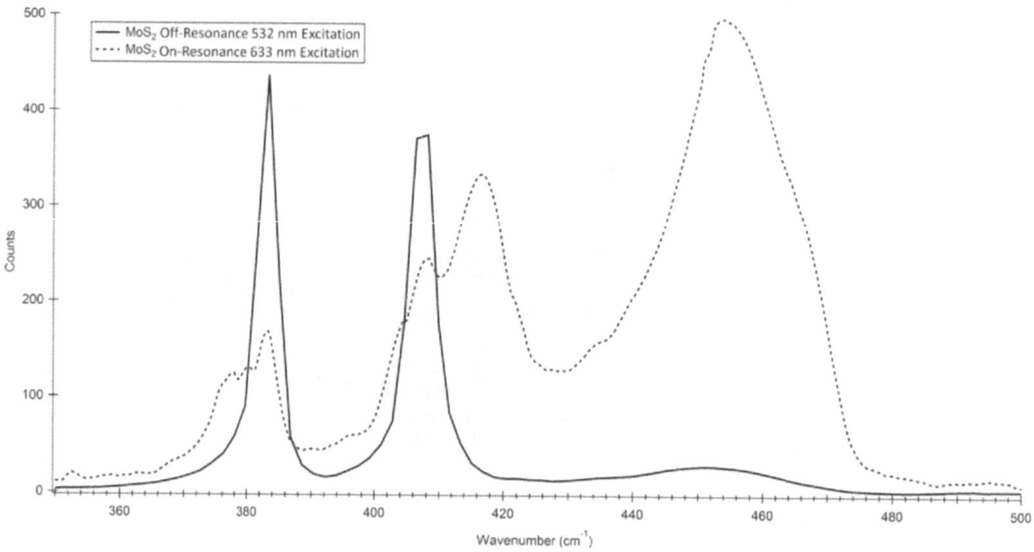

Figure 3.38

Resonance Raman (633 nm excitation) and off-resonance Raman (532 nm excitation) of MoS_2. The peaks that were too weak to see with the off-resonance 532 nm excitation are enhanced with the 633 nm resonance Raman excitation.

of fluorescence emission occurring is also enhanced when the excitation light is in resonance with an electronic transition.

For resonance Raman experiments to be a success, you often simply have to be lucky in choosing a sample for which the excitation laser lines on your Raman system are in resonance with an electronic transition in your sample and that the fluorescence from the sample does not overwhelm the Raman signal (or work with a sample that intrinsically does not have strong fluorescence).

3.5.13 Coherent Anti-Stokes Raman Spectroscopy

Biological samples tend to be hydrated and therefore tend to auto-fluoresce (meaning fluoresce even without fluorescent labels) strongly. This makes them a challenge for FTIR and Raman. One way to deal with these issues is to move to anti-Stokes Raman scattering. The anti-Stokes Raman scattering signal will be at a higher energy than the excitation light. The fluorescence is always at a lower energy than the excitation light. This means that it is easy to spectrally separate the anti-Stokes Raman signal from the auto-fluorescence of the sample. The challenge in this case is that the anti-Stokes Raman signal is orders of magnitude weaker than the Stokes Raman scattering signal, which was quite weak to start with.

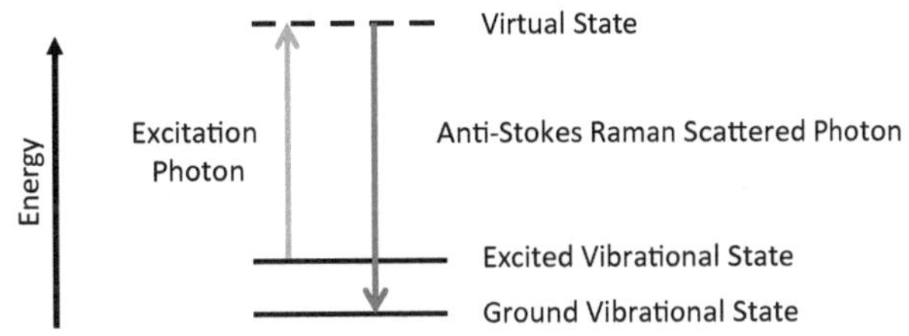

Figure 3.39

Energy level diagram of anti-Stokes Raman scattering. The molecule starts in an excited vibrational state and transitions down to the ground state, scattering a photon with more energy than the excitation photon.

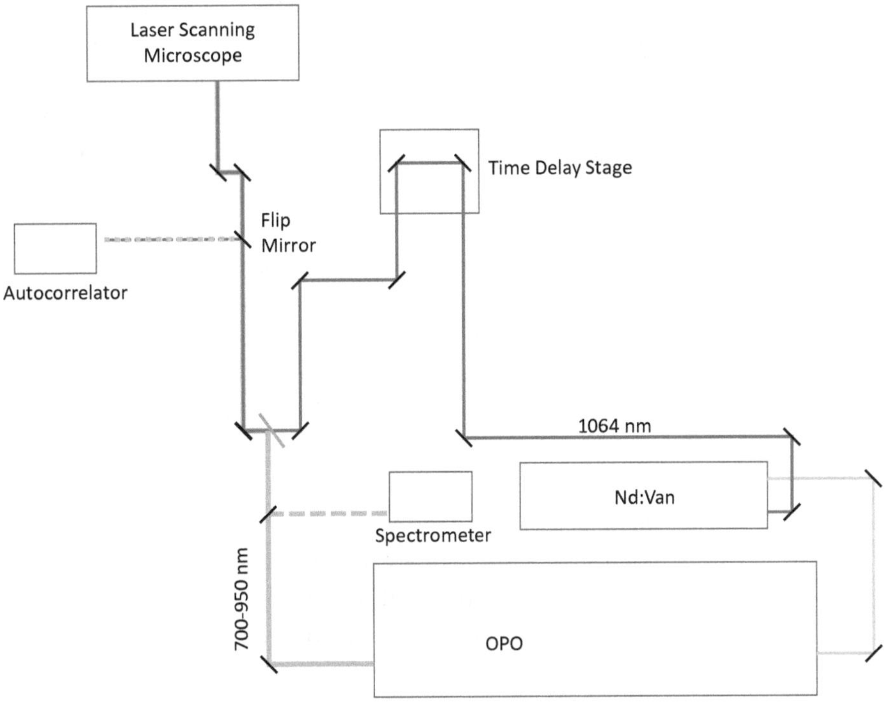

Figure 3.40

Schematic of coherent anti-Stokes Raman scattering setup.

The anti-Stokes Raman shift needs a molecule to already be at an excited vibrational energy level to start. At room temperature, there are relatively few molecules sitting in an excited vibrational state.

Coherent anti-Stokes Raman scattering (CARS) offers a solution to this problem. In CARS, spectroscopy begins by stimulating a number of molecules into an excited vibrational state with some extra lasers, and then proceeding with anti-Stokes Raman scattering. This is performed with two extra laser beams. The energy difference between these two laser beams is tuned to the difference in energy levels that you are trying to populate. The first two laser beams move the molecule of interest into the excited vibrational state, and then a third photon causes an anti-Stokes Raman scattering event. The two excitation beams are referred to as the pump beam, usually a fixed wavelength, and the Stokes beam, usually a tunable wavelength from a synchronously pumped optical parametric oscillator (Section 2.2.7). The pump beam often also provides the third photon, so in practice only two synchronized pulsed laser beams are needed.

The laser beams for CARS need to be pulsed lasers. For narrowband CARS (you are only exciting a single vibrational mode), picosecond lasers are used. For broadband CARS (you excite multiple excitations simultaneously), femtosecond lasers are used.

One advantage of CARS is that it is a nonlinear multiphoton process, so signal is only generated from the focal point. This allows for three-dimensional images to be built up by scanning the laser focal spot through x, y, and z of the sample. (Further explanation of multiphoton microscopy is given in Section 4.11.)

Two downsides of CARS are that there is often a large non-resonant background to the signal that decreases contrast, and that you can only scan a single vibrational mode in narrowband CARS or a narrow range of vibrational modes in the case of broadband CARS. Tuning the laser repeatedly to build up a full CARS spectrum would take a prohibitively long time. Since CARS is limited in the spectral range it can collect, it loses the ability to take advantage of chemometric techniques that help deal with the complexity of the overlapping signals in biological samples.

Currently available laser technology also limits how closely you can tune the pump and Stokes laser beams to $>900\,\text{cm}^{-1}$. Additionally, there are clearly many interesting and important vibrational modes at lower frequencies that cannot be accessed with CARS at this time.

There are a number of variations of CARS microscopy that attempt to deal with the non-resonant background issue. CARS spectroscopy systems are usually homebuilt or custom built setups.

3.5.14 Stimulated Raman Spectroscopy

Stimulated Raman scattering (SRS) is one approach to dealing with the non-resonant background that plagued CARS imaging in the previous section. In SRS, you carefully count the number of excitation photons instead of recording the number of scattered photons. If you successfully move a molecule from the ground state to an excited vibrational state, the pump beam loses a photon and the

Stokes beam gains a photon. Stimulated Raman scattering requires very stable lasers. An SRS setup incorporates an electro optic modulator to modulate the beams (Section 2.3.54) and a lock-in amplifier to only record signals produced at the same frequency of the beam modulation. SRS imaging offers the chemical specificity of CARS and the three-dimensional imaging capabilities of other multiphoton microscopy techniques (Section 4.11). However, it has the drawback that SRS can only examine a single peak at a time, and thus cannot easily take advantage of multivariate analysis techniques.

3.5.15 Surface Enhanced Raman Spectroscopy

When light interacts with a sharp conductive tip, the electric field will be enhanced. This is the so-called "lightning rod effect." If a sample is spread out on a surface with many sharp conductive tips, the molecules sitting on the tip will have an enhanced probability of Raman scattering. This is referred to as surface enhanced Raman spectroscopy (SERS).

Surface enhanced Raman spectroscopy can be performed on any conventional Raman system and is really a sample preparation technique. By purchasing or preparing a metal substrate (usually gold or silver) with many sharp tips and then spreading the sample of interest on the substrate, the signal from each molecule sitting on a sharp metal tip will be enhanced. The exact enhancement factor is still a matter of debate, with many different values reported in the literature. Generally, the enhancement depends on the geometry and material of the metal protrusions, as well as the nature of the interaction between the molecule of interest and the metal protrusions. The discrepancy in the literature probably stems from the wide variation of the SERS substrates produced. In any case, when SERS samples are prepared well, one can achieve an enhancement factor of several orders of magnitude to Raman scattering signals.

One of the big challenges of SERS is preventing contamination of the SERS substrates. There are enough random molecules in the atmosphere (perfume, sneezing, exhaust from a car, etc.) that can occupy surface sites on the substrate that you really wanted your sample molecules to occupy. Many times, without proper cleaning and careful handling of SERS substrates, no enhancement will be observed due to contamination.

3.5.16 Tip Enhanced Raman Spectroscopy

An enhancement effect can also be observed at a single point with a sharp, metal-coated atomic force microscope (AFM) tip (Section 4.18). This is referred to as tip enhanced Raman spectroscopy (TERS). The advantage is that you can now bring the metal tip to the sample rather than having to spread the sample over an array of metal tips. Tip enhanced Raman spectroscopy also allows you to localize the area that you are collecting Raman spectra from. The spatial resolution of TERS imaging is determined by the radius of the AFM tip used (~10–20 nm). Keeping the Raman excitation laser aligned with the AFM tip is a challenge, and a number of setup geometries have been implemented (side illumination, bottom illumination, etc.)

In theory, TERS images can be generated by scanning the tip over the sample and collecting a Raman spectrum at each point. In practice, the data collection tends to be prohibitively slow and the AFM systems are not stable over the long time periods that would be needed to collect a TERS image. So usually a highly localized Raman spectrum is collected from a point of interest as determined by the AFM image. Tip enhanced Raman spectroscopy tips are also plagued by fouling issues. If any part of your sample starts to stick to your tip, then you must start over with a clean AFM tip.

3.6 Laser Induced Breakdown Spectroscopy

Laser induced breakdown spectroscopy (LIBS) is one of the analytical techniques that was included on the Mars Rover Curiosity's ChemCam system. Curiosity used LIBS on Mars to help provide chemical information about the Martian geology. Laser induced breakdown spectroscopy literally zaps a rock with a laser to vaporize a small amount of material, and then analyzes the emitted light with a spectrometer to determine the atomic composition.

Underlying Physical Principles

Laser induced breakdown spectroscopy is an atomic emission spectroscopy technique. It uses a high-intensity laser pulse to form a plasma, and then measures the characteristic atomic emission lines from the plasma. If the electric field of the laser pulse is strong enough, the electrons forming the chemical bonds in the sample can be ripped away and the material in the sample falls apart. The electrons in the plasma will eventually relax back into atomic orbits and emit a photon at a characteristic energy (e.g., Sections 3.2 and 3.3). Since each element has a unique set of electron energy levels, the atomic composition of the sample can be determined by analyzing the energies of light that are emitted from the sample as it is broken down by the laser pulse.

Laser induced breakdown spectroscopy can be performed with nanosecond, picosecond, or femtosecond pulses. The shorter the pulse, the less thermal damage is done to the sample area surrounding the ablated spot. The detector is often gated, which means it only collects light after it receives a signal that the laser pulse is occurring. Gated detectors help to suppress background light.

What Scientific Questions Can Be Asked?

Laser induced breakdown spectroscopy determines the atomic compositions of samples. The interaction volume of LIBS can be small enough that it is even used in the art conservation field to determine the composition of paints on delicate works of art.

Strengths and Limitations

By nature, LIBS is a destructive analytical method. Granted, LIBS can be performed on a very small area on the sample, but it only provides information about the area that was destroyed. If your sample is inhomogeneous, then you will need to destroy a larger area to be sure that you are collecting a representative data set of the atomic composition of the sample.

What Samples Are Appropriate?

Laser induced breakdown spectroscopy can be applied from a distance, or can be fiber coupled with the same fiber optic, providing laser excitation to the sample and collecting emission spectra from the sample. For these reasons, compositional analysis using LIBS can be performed without complicated sample mounting or preparation.

Common Pitfalls

Fluorescent ceiling lights contain a plasma that also produces sharp emission lines. Be sure to run negative control experiments where you analyze a blank sample to be sure you are not collecting the spectrum of the overhead fluorescent lights. Using a gated detector helps reduce the contribution of background light to the signal.

Atomic emission lines of interest are sometimes in the UV spectral range. If you are building your own LIBS system, double-check that the optical components you are using are acceptable for use in the UV region. A standard silicon-based CCD camera is typically sensitive down to 200 nm, but glass lenses, mirrors, diffraction gratings in the spectrometer, etc. need to be specially selected to be compatible with the UV light.

Schematics

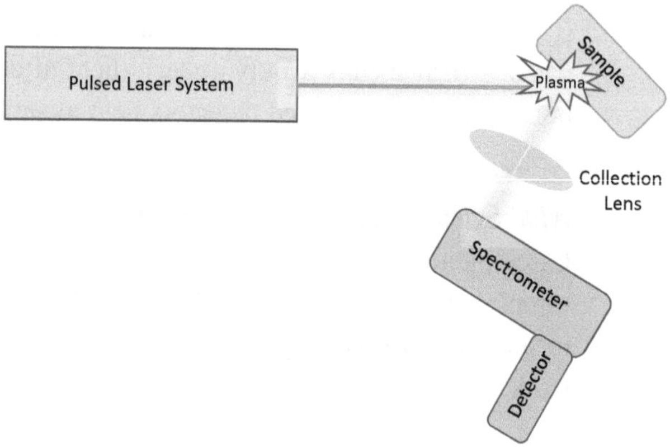

Figure 3.41

Schematic of a LIBS setup.

Sample Data and Data Interpretation

The National Institute of Standards and Technology has an extensive database of atomic spectroscopic data (www.nist.gov/pml/atomic-spectroscopy-databases).

3.7 Hyperspectral Imaging/Mapping

Underlying Physical Principles

Hyperspectral imaging means that there is a spectrum across some wavelength range for each pixel in an image. (The term "hyperspectral" is a bit of a fuzzy buzz word.) "Hyper" comes from mathematics, and refers to objects with more than three dimensions that are hard for humans to visualize. In a hyperspectral image, each pixel has an x and y coordinate and intensity values at multiple wavelengths. Hyperspectral imaging tends to be used in the context of where there is not a single physical phenomenon corresponding to the spectral information. Raman mapping and FTIR mapping are technically hyperspectral images, but are not usually referred to as such; UV-VIS-NIR hyperspectral imaging is the most common. Often, the spectrum associated with each pixel is a convolution of absorption, reflection, transmission, and scattering phenomena all mixed together.

A hyperspectral image tends to be an image at different wavelengths in the visible or near IR looking at reflection or scattering of light. Hyperspectral imaging is often used for macroscopic and microscopic samples. For example, the Landsat program by the United States Geological Survey utilizes hyperspectral

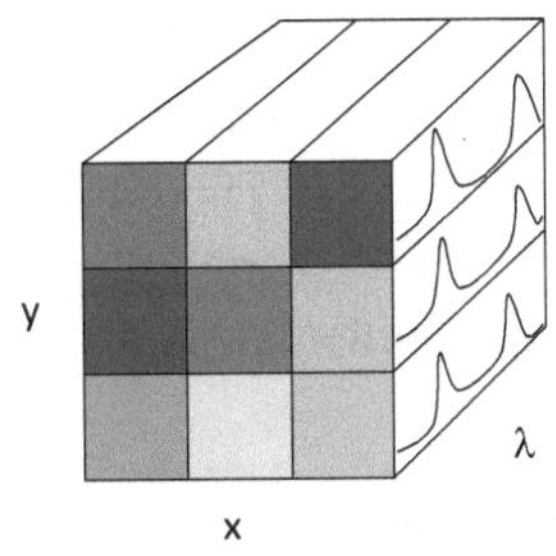

Figure 3.42

A hyperspectral data set contains a full spectrum for each *x*, *y* coordinate set in the image. The intensity of the pixel is the representation of some quality of interest of the spectrum (e.g., peak height at a certain wavelength).

imaging. Satellites take images of Earth at several wavelengths ranging from UV to near IR. The images capture the reflectivity of different geographic features at different wavelengths, and while it is difficult to say exactly what combination of reflectivity, absorbance, and scattering phenomena are contributing to the contrast in the image, there are clear features that can be analyzed for information. For instance, hyperspectral imaging has been widely applied to agriculture to remotely determine crop health and yield.

What Scientific Questions Can Be Asked?

Spatial distribution or variation of one or more properties of a sample can be rapidly determined with hyperspectral imaging techniques.

Strengths and Limitations

The hyperspectral map itself is less useful without extra information about the sample, such as what you suspect the sample is made of. Hyperspectral imaging can be used to map quickly compared to FTIR or Raman mapping.

Data handling and visualization in hyperspectral imaging are challenging. For instance, plotting data is challenging because we can only represent a limited subset of the hyperspectral data. Each pixel in a standard color three-dimensional image can have three spatial variables (x, y, z) and then three other variables encoded in the intensities of the red, green, and blue channels.

Because a full spectrum is associated with each pixel, the data sets also become large quickly, so data and file management can pose problems. Also, so much information is collected that you will need to start relying on multivariate analysis for data processing and interpretation.

What Samples Are Appropriate?

Depending on the optical setup, nearly any sample could be imaged with hyperspectral techniques. The technique has been applied to everything from agricultural fields with aerial photography to nanoparticles in biological cells in microscope configurations.

Common Pitfalls

The pitfalls of hyperspectral imaging will be highly sample- and experiment-dependent. In general, it is extra important to perform positive and negative controls, as the spectral features tend to be much broader and matrix effects can play a larger role in the spectra features compared to FTIR or Raman mapping.

Schematics

For the aerial geographic hyperspectral projects, the images are often taken with different optical filters, so the wavelength resolution is limited to just a couple wavelength points (e.g., UV, VIS, NIR, SWIR).

For microscopic hyperspectral techniques, broadband white light is used to illuminate the sample. The systems are often built as a line scan system where the slit of the spectrometer is used as a spatial filter to collect the spectra from a vertical line of the sample, and then the sample stage is moved to a new line to build up a hyperspectral image.

Sample Data and Data Interpretation

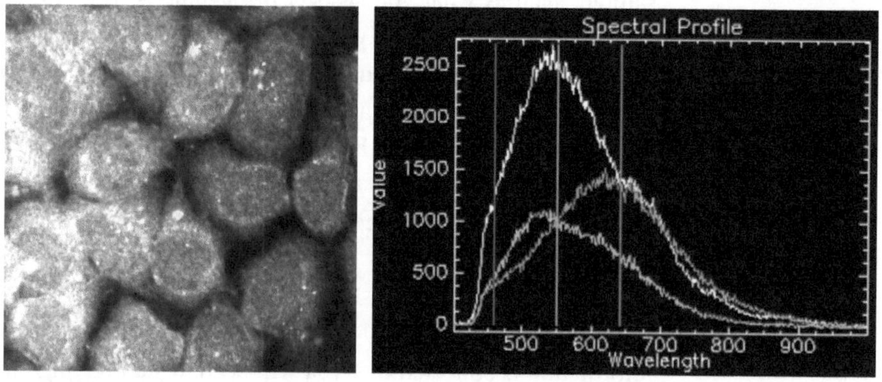

Figure 3.43

Sample hyperspectral image with several representative spectra from different individual pixels.

3.8 Ultrafast Pump–Probe Time-Resolved Measurements

Samples in excited electronic states are highly dynamic systems, so it is natural to ask what is happening as a function of time. One way to investigate systems on very short time scales is to use an ultrafast laser in a pump–probe spectroscopy configuration. Here, one laser pulse (the pump pulse) puts the sample into an excited electronic state and a second laser pulse (the probe pulse) spectroscopically interrogates the system at a controlled time later. This is repeated at a series of different time delays between the pump pulse and the probed pulse to build up a data set about the dynamics of the excited state system.

While there are some commercial time correlated single photon counting (TCSPC) systems that can be purchased off the shelf, pump–probe setups are almost exclusively homebuilt or custom built. TCSPC will reliably provide dynamics on the time scale of tens of picoseconds to nanoseconds. If you are interested in sub-picosecond dynamics, you will have to build a pump–probe optical setup.

Underlying Physical Principles

An "ultrafast" laser has pulse lengths of 100 femtoseconds or less. A femtosecond is 10^{-15} seconds. On the femtosecond time scale, electrons can react to the electric field in the laser pulse, but the atomic nuclei cannot move fast enough to respond to the electric field given their larger mass. So, ultrafast spectroscopy allows the decoupling of electron and nuclear dynamics.

In pump–probe spectroscopy, the output of a single ultrafast laser is split into two beams, often with unequal intensity with a beam splitter. (Splitting the output of a single laser is the easiest way to ensure that the laser pulses are synchronized in time. The downside of this approach is that it limits the wavelength differences between the two laser pulses. Synchronously pumped optical parametric oscillators give a bit more flexibility on the choice of wavelengths, but are more expensive.) The two synchronized laser beams are routed around the optical table, making sure that the optical path length is exactly the same for both beams.

One laser beam is denoted the pump beam and the other is denoted the probe beam. The pump beam typically has a higher intensity than the probe beam. The pump pulse will excite electrons in the sample. The probe pulse will arrive at a controlled time later and probe the state of the electrons in the sample. There is the assumption that the probe beam is weak enough not to cause too much perturbation to the sample with its interaction.

To control the timing between the pump pulse and the probe pulse, the probe optical path typically has an optical delay line, which consists of two mirrors on a mechanical translation stage. By moving the delay line, the time at which the probe pulse arrives at the sample relative to the pump pulse can be controlled without the need for ultra-high speed electronics.

The assumption is that your sample returns to the ground state between each set of pulses and that your sample is stable and can be excited and probed many times. With a 76 MHz repetition rate laser, the time between pump pulses is ~13 ns. This pump–probe technique also allows you to use slow detectors such as photo multiplier tubes and then signal average over multiple pump–probe pulse pairs.

TIP: The easiest way to coarsely verify that the optical paths are the same length in an optical delay line is to use a piece of string to measure each of the optical paths. Compare the lengths of the two strings and then adjust the optical paths as needed. This will get you surprisingly close to matched optical paths. For fine adjustments, you will need to scan the optical delay line and monitor the sum frequency signal from a reference sample such as z-cut quartz that has a strong sum frequency generation signal. The sum frequency signal will only occur when both laser pulses interact with the sample in the same time and space. To check that you have a sum frequency signal, first block one beam and then the other. The sum frequency signal will disappear when either beam is blocked. If your pump and probe beams are coming into the sample at different angles, the second harmonic signal generated from the sample will be co-linear with the excitation beams, but the sum frequency signal will be spatially separated between the two excitation beams.

The pump beam puts the sample into an excited state. When the probe beam interacts with the sample, several different phenomena can occur. The electron can be excited to an even higher electronic state, leading to increased absorption (a decrease in intensity) of the probe beam; the electron can be stimulated to emit, leading to an increase in intensity of the probe beam to a higher level than if the sample were not present; or the ground state electrons can become depleted to the point that there are no more electrons for the probe beam to excite, so the probe beam transmission increases to the level of no sample being present.

When running a pump–probe experiment it is important to include some negative time points (time points where the probe beam arrives before the pump beam), as these provide a baseline for your measurement. There is often a bit of messiness that is hard to understand when both the pump and the probe beams are simultaneously interacting with the sample, but as soon as the two pulses are moved apart in time by one full pulse length, the data should become more interpretable.

What Scientific Questions Can Be Asked?

Electrons in an excited state will have different interactions with the probe beam than electrons in the ground state. Factors such as changes in the absorption (transient absorption spectroscopy) or reflection (transient reflection spectroscopy) or second harmonic generation (time-resolved SHG) of the material can be probed and correlated to electron relaxation rates, carrier lifetimes, etc. The time that the electron stays in the excited state is very sensitive to the exact chemical composition of the sample, to the local chemical environment, and to defects in the sample. For instance, pump–probe spectroscopy is often applied to characterize carrier lifetimes in semiconductors under different processing conditions.

Strengths and Limitations

Pump–probe techniques allow you to probe the electron dynamics of a sample on a sub-picosecond time scale. This can be quite informative about the local chemical environment in the sample. Data interpretation can be challenging, however, for some of these pump–probe experiments. You can get an electron relaxation rate out of the data, but what does it really mean? Why is the rate what it is and how can you control it?

What Samples Are Appropriate?

Pump–probe techniques have been applied to a wide range of samples including biomolecules, semiconductors, nanomaterials, and even historic pigments in paintings. The challenge is really in matching the pump and probe laser wavelengths to the sample. You need to be able to excite electrons in the sample without damaging the sample, and to monitor a change in the response of the excited electrons to the probe pulse by measuring some sort of signal, such as absorption, reflection, fluorescence, photoluminescence, second harmonic generation, etc.

Common Pitfalls

Keeping the pump and the probe pulses aligned in space as you translate the optical delay line is one of the challenges in pump–probe techniques. If the overlap in space of the pump and the probe beam changes with the translation of the optical delay line, you will measure a shorter relaxation rate than is really occurring in the sample. The best way to guard against this is to find a couple reference samples with different known relaxation rates and check your optical setup. If each sample yields about the same decay rate, you are really measuring the **spatial walk off** of the probe beam from the sample volume that was excited by the pump beam.

In order to improve the signal-to-noise ratio in ultrafast measurements, the intensity of the pump beam is often modulated with an electro optic modulator (Section 2.3.54) or an acoustic optical modulator (Section 2.3.53) and a lock-in amplifier is used on the detection side to only look for signal that is at the same frequency as the modulation of the pump beam.

Schematics

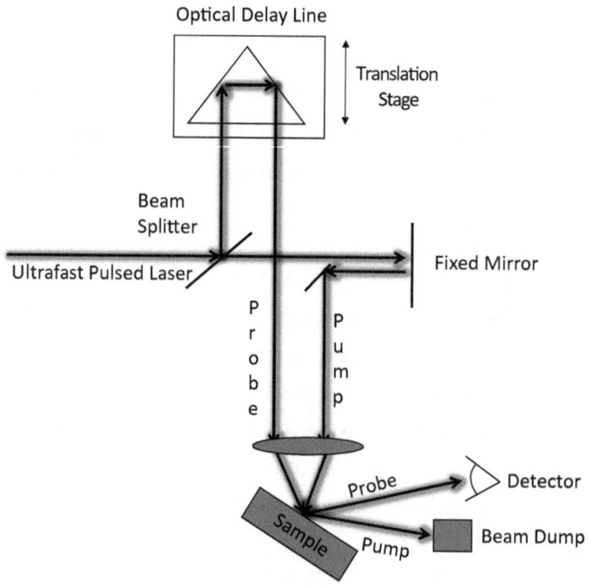

Figure 3.44

A simplified schematic of an ultrafast pump-probe optical setup for transient reflection measurements. The detector does not have to be fast, since the time-resolved portion of the measurement is controlled by the time delay between the pump and the probe beam, which is determined by the position of the optical delay line.

Sample Data and Data Interpretation

The data from a pump–probe experiment typically lie on an exponential or bi-exponential decay curve. These can be directly fitted with a multi-exponential decay, but the accuracy can be quite low. You can fit almost anything with enough exponential components. Phasor analysis is a more complicated but more robust analysis that is independent of *a-priori* assumptions (such as how many exponential components to use) to give you lifetimes for the different components. Phasor analysis works by taking a Fourier transform of the time-resolved curve and mapping the real and imaginary parts of the signal against each other. If there is a single exponential component in your data, the plotted point will land on what is called the universal semi-circle. If your data do not land on the universal semi-circle, then you have multiple exponential components to your data. The exact values can be read off the universal semicircle plot. A full explanation of phasor analysis is beyond the scope of this book, but the authors encourage those interested in time-resolved spectroscopy to strongly consider phasor analysis as

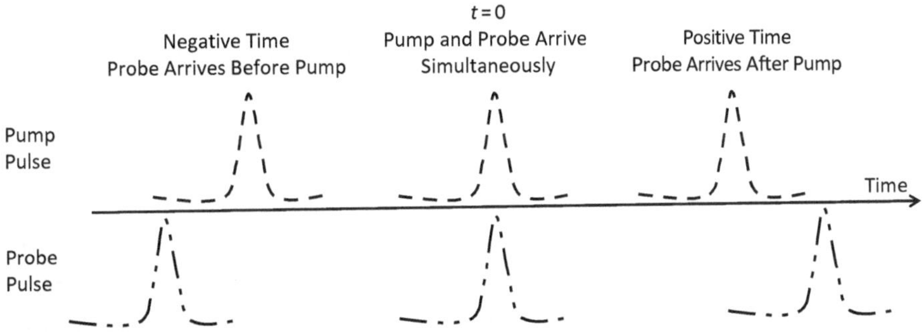

Figure 3.45

The time resolution of the measurements is determined by the relative time of the pump and probe pulse arrivals at the sample. Negative times correspond to the probe arriving before the pump. Time zero corresponds to the instant when the two pulses arrive simultaneously. Positive time corresponds to the probe arriving after the pump.

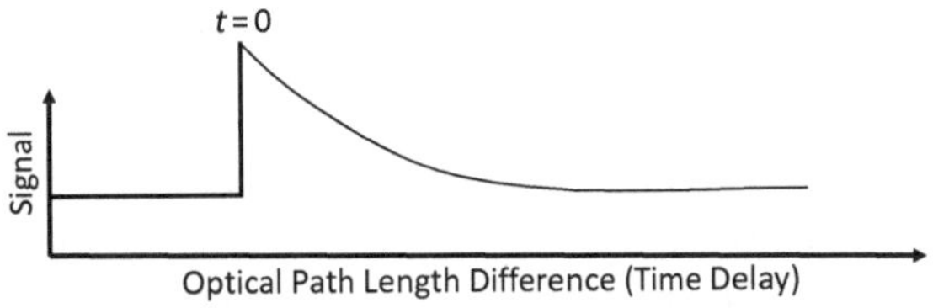

Figure 3.46

Pump–probe data typically look like a multi-exponential decay after time zero.

a much more robust method of time-resolved spectroscopy data analysis and fitting compared to multi-exponential fitting.

3.9 Chemometrics: Multivariate Data Analysis

All too often, spectroscopists focus on a single peak for data interpretation and ignore all the other peaks. This of course means that we are throwing away a lot of information that we worked hard to collect about our samples, simply because it is hard for us to interpret and recognize patterns in it. Wouldn't it be great if we could use all the information that we collected about the sample, maybe even across different techniques?

Chemometrics is a collection of data-processing techniques that applies linear algebra and multivariate analysis techniques to spectroscopic data. The techniques are borrowed heavily from the social sciences, where large, messy data sets have traditionally been the norm. With the increasing automation of instrumentation in the physical, natural, and life sciences, it is becoming easy to generate huge amounts of data that are physically impossible for a human to comprehend without computational assistance.

One point to make about how to think about your data for chemometrics is that a spectrum is not a single measurement. Mathematically, a spectrum is actually hundreds of measurements. An FTIR spectrum consists of the IR absorbance of the sample at each wavenumber between 4000 and 600 cm^{-1}. Some of these measurements are strongly correlated. Of course, the measurements at 1705 cm^{-1} and 1706 cm^{-1} will be correlated, but the measurement at 1705 cm^{-1} (the C=O stretch) is independent of the measurement at 2850 cm^{-1} (the CH_2 stretch). There may be systematic variations between multiple peaks that are too subtle for humans to notice but are easy for pattern recognition algorithms to pull out.

Chemometric data analysis is performed using software packages. Typically for chemometric software packages, you organize your data set into a matrix with samples as rows and measurements as columns. You can even combine data from multiple techniques into the matrix. In a public health study, for instance, you would have a number of different measurements in a variety of units (height, weight, blood pressure, blood oxygen levels, etc.). This works because chemometric algorithms are looking for the systematic variance between multiple measurements and do not care what technique the data were acquired with or what the units of the measurements are. Chemometrics can take all the information you have about the samples, despite the fact that it was collected from multiple techniques, and help you see the relationships across a population of samples.

3.9.1 Principal Component Analysis

Principal component analysis (PCA) allows one to change the basis set used to describe the data to one that is more natural for the data. A rough example of when changing basis sets makes sense is giving directions in a city. If the streets do not run north–south, then giving directions in terms of north, south, east, and west doesn't make much sense. Instead, giving directions along the streets makes more sense. Consider Figure 3.47. Giving directions from the arrow to the star, one would naturally say, "Go two blocks then turn left for another two blocks," instead of suggesting, "Go north 800 meters and east 200 meters." The two arrows showing the directions along the street would be referred to as principal component 1 (PC1) and principal component 2 (PC2) in chemometric jargon.

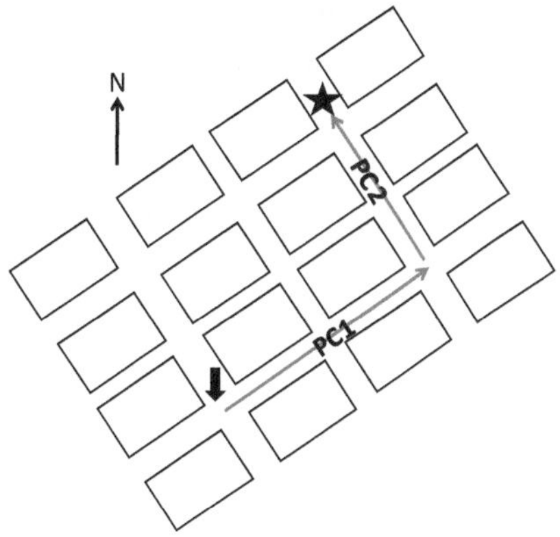

Figure 3.47

An example of when changing basis set makes sense to describe your data.

The powerful part about PCA is that you as a human do not need to know what the new basis set to describe your spectra is. A PCA algorithm can crunch through the data set for you and provide a new basis set that best describes how different samples in your data set are similar or different. PCA does a terrific job of determining how the spectra in a data set are related to each other. Principal component analysis does need multiple measurements to give a robust answer. Usually the more samples measured, the better the PCA model.

3.9.2 Data Preprocessing

A variety of preprocessing techniques can help highlight the important variance in a data set and suppress the unimportant variance in a data set. Beware, however: the old computer programming adage of "garbage in = garbage out" holds when doing multivariate analysis of spectroscopic data sets.

For this reason, be sure to do an appropriate selection of spectral regions where the variance in the measurement is from your sample and significant to your scientific question. For example, in FTIR spectra, not including the CO_2 spectral region (around $2350\,cm^{-1}$) is a good idea as the variations in the CO_2 peaks are usually not from your sample, but from the instrument operator changing how much she or he is breathing near the instrument. For UV-VIS-NIR absorbance, leaving out the small spectral regions around the source change, grating change, and detector change is probably a good idea, as the variance there is from the instrument rather than your sample.

Mean centering is a preprocessing technique that works well for spectral data sets. Mean centering averages all the spectra together and then subtracts the mean spectra from each of the individual spectra. This highlights the systematic differences in the spectra from each other.

Normalizing your spectral data can help or can mask important variance, depending on your experimental setup. Think about whether you are looking for new chemical bonds (shifts on the x-axis) or changes in concentrations (shifts along the y-axis).

3.9.3 Classical Least Squares

A classical least squares (CLS) approach is used to decompose a spectrum of a mixed sample into the individual component spectra. For example, let's say you are recreating a coating on a sample and you measure the spectrum of the composite coating. You know that the coating is a mixture of materials and want to find the relative amounts of the different materials in the mix. To extract the relative composition of the coating, first collect the spectra of the pure reference components (and maybe spectra of some recreated mixes if the components reacted with each other when they were mixed). Then, with a classical least squares model you could predict that the mixture is 30 percent compound A and 70 percent compound B.

3.9.4 Partial Least Squares

Partial least squares (PLS) is a technique used to make predictions about unknown samples based on their measured properties. Abstractly, if you have measured x, you can use PLS to predict y.

As a concrete example, say you have a data set of near IR absorbance spectra of different beers and the alcohol contents of each of those beers. Using these data, your goal is to build a model to predict the alcohol content of a new beer based on the near IR spectrum measured. You can use a PLS model to make such predictions. As another example, suppose you have a data set of FTIR absorbance spectra for clays fired to different firing temperatures and you want to estimate the firing temperature of an ancient clay pot based on the FTIR spectra. Building a PLS model can leverage the data you already have about known samples to inform your conclusions about a new, unknown sample.

For PLS modeling, you need calibration and test data sets to build up a reliable model and make accurate predictions, but such predictive modeling can be quite powerful where optical spectroscopy techniques are faster and easier measurements to make than some of the other traditional measurement techniques.

3.9.5 Clustering

Clustering measures how closely samples reside in multivariate space and then makes a dendrogram to highlight relations between samples in a data set. For instance, clustering can inform which samples are most similar or the most different from each other.

3.9.6 Software Packages

There are a variety of software packages for such data analysis. The PLS Toolbox from EigenVector Research Inc. provides a graphical user interface to run on top of MatLab. The Unscrambler by Cameo is another similar package. Both of these work well for spectral data sets. MatLab itself has some PCA algorithms built in. The geospatial community uses ENVI to process hyperspectral data sets. The social science and health science communities have their own favorite multivariate analysis packages.

3.10 Further Reading About Spectroscopy Techniques

Spectroscopyonline.com, while quite commercial in its orientation, does have a lot of practical advice columns by scientists who do lab work.

Fundamentals of Analytical Chemistry by Skoog and Haller is an undergraduate analytical chemistry text that does a good job of describing many of these techniques from an analytical chemistry perspective.

Physical Chemistry by Peter Atkins is an undergraduate physical chemistry book that gives a lot of nice, very clear explanations of many of these spectroscopic techniques from a physical chemistry perspective.

Molecular Spectroscopy by Jeanne McHale is a graduate-level text from a chemistry perspective that delves deeper into the mathematics describing the spectroscopic techniques.

Principles of Fluorescence Spectroscopy by Joseph Lakowicz is an excellent resource on all the details of fluorescence spectroscopy.

The National Institute of Standards and Technology (NIST) has an extensive database for the interpretation of FTIR, Raman, UV-VIS-NIR, LIBS, and other atomic emission techniques (www.nist.gov/pml/atomic-spectroscopy-databases and http://webbook.nist.gov/chemistry).

For further information on nonlinear spectroscopy, *Nonlinear Optics* by Robert Boyd and *The Principles of Nonlinear Optics* by Y.R. Ron Shen both have excellent texts, but written at a graduate physics level.

TCSPC Handbook by Becker & Hickl provides a lot of detailed background information about TCSPC techniques and how to implement them on a variety of microscopes (with Becker & Hickl TCSPC detectors and software, of course).

Chemical Applications of Group Theory by Albert Cotton is a go-to for physical chemists interested in vibrational (Raman and FTIR) and electronic spectroscopies (fluorescence and photoluminescence), and their associated selection rules.

4 Optical Imaging: What Are the Pretty Pictures Actually Showing Me?

There are many optical imaging modalities, each of which uses different physical phenomena to generate contrast in an image. Entire books have been devoted to many of these techniques individually. This chapter introduces you to the most common modalities in use today and gives a feel for the underlying physical principles behind each technique so that you can choose the most appropriate one for your sample and scientific question. We encourage you to seek out the more in-depth books once you have identified which technique or modality you are interested in.

4.1 A Quick Tour of an Optical Microscope

Most users can identify the eyepiece, sample stage, and focus knobs on a microscope, but are unaware of the functions of the plethora of other knobs available for adjustment on standard laboratory microscopy. In this section, we provide a guided tour of the optical path of a research-grade optical microscope so that you will be able to identify all the knobs and know how to adjust them properly for optimal image collection.

Sorting out which way is up is a good first step when you approach an unfamiliar microscope. If the microscope is an upright microscope, the microscope objective lens will be above the sample stage. If the microscope is an inverted microscope, the objective lens will be below the microscope stage. Then, begin tracing the light path to understand the microscope setup.

Let us begin our tour of the optical path at the beginning, with the light source. There are many different light sources for a microscope. The classic white light source is a halogen bulb, which contains a metal filament heated by an electric current, making it glow white hot. The light from these bulbs is fairly gentle on samples and sufficient for most transmitted light techniques such as bright field imaging, dark field imaging, phase contrast imaging, and differential interference contrast imaging (DIC). For fluorescent imaging techniques, brighter white light sources, such as mercury lamps or xenon lamps in combination with filters, can be

Figure 4.1

A research-grade microscope is typically a confusing and chaotic collection of unlabeled, poorly labeled, or misleadingly labeled knobs, wheels, and push rods. Taking time to trace out the optical path will help you identify all the components and enable you to make the best use of the instrument.

used. Light emitting diodes and lasers are also widely used in fluorescent imaging techniques.

The light source is typically fed into the microscope body at the back of the frame. A microscope can have multiple light sources to enable multiple techniques on a single microscope body, including multiple ports for white light sources and lasers. It is worthwhile taking a moment to poke around the back of the microscope body when you first walk up to an unfamiliar microscope to assess what light sources are available and where they enter the microscope body.

A lens immediately follows the light source to collimate the light. A diffuser also usually follows halogen bulbs. Diffusers are pieces of ground glass that spread out the light and provide even illumination so that the image of the bulb's filament is not in your data.

A shutter is usually placed directly after the light source. A shutter typically consists of a small piece of black metal that can be moved in and out of the beam path. The control for the shutter can be in a variety of places. Sometimes there is a simple mechanical slider or push/pull rod. Sometimes there is a push button remote. Sometimes the shutter control is hidden in the software that is controlling the microscope system. Understanding how to turn on the power supply to the

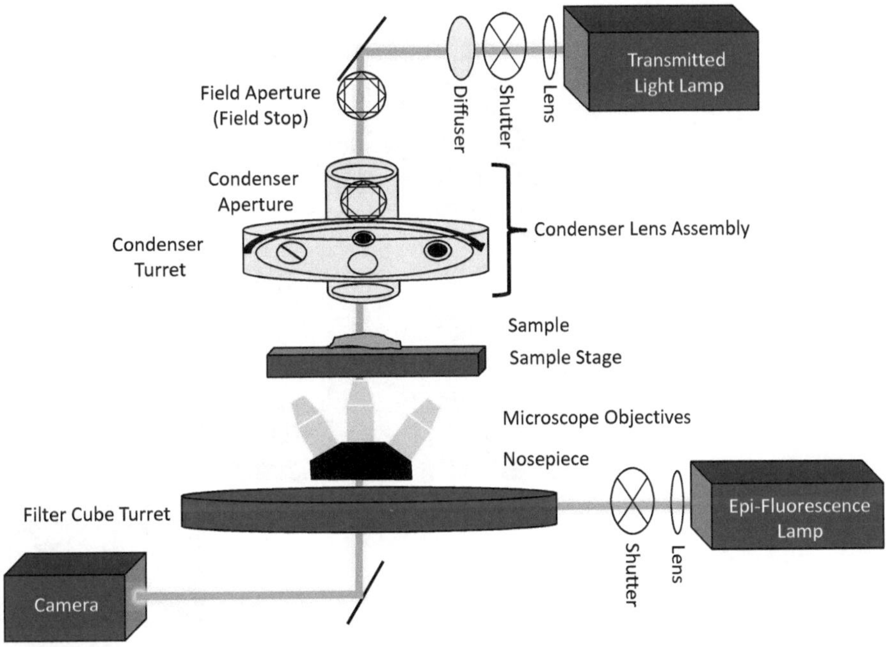

Figure 4.2

Schematic of an optical microscope. Even in a simplified cartoon form, a microscope has many parts to understand. A good practice is to follow the light path to understand what is going on.

light source and how to open and close the shutter are important first steps when approaching an unfamiliar microscope system.

The field aperture (sometimes called the field diaphragm or field stop) is the next element in the optical path. It is used during alignment of the condenser assembly for Köhler illumination for transmitted light techniques (see Section 4.1.1). It is *not* for adjusting the intensity of the incident light. There is either a series of neutral density filters or a voltage control dial to control the light intensity. Your image quality will be degraded if you use the field aperture to adjust light intensity.

For transmitted light techniques, the next stop on the tour is the condenser lens assembly. Proper adjustment of the condenser lens assembly is essential for good results from transmitted light techniques. The proper adjustment of the condenser assembly ensures even illumination of the sample. There are many unmarked knobs on the condenser assembly that must be properly adjusted for optimal imaging. See Figure 4.8.

For all transmitted light techniques, the lens in the condenser needs to be focused and centered. This is accomplished by closing the field aperture and bringing the image of the edge of the field aperture into focus with the focusing

(a) (b)

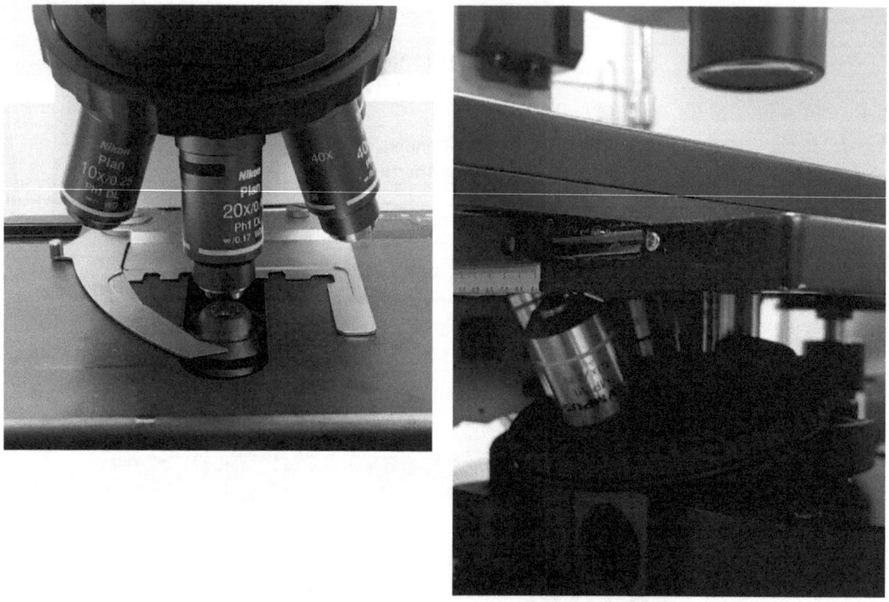

Figure 4.3

(a) Upright microscope, in which the objectives are above the sample stage. (b) Inverted microscope, in which the objectives are below the sample stage.

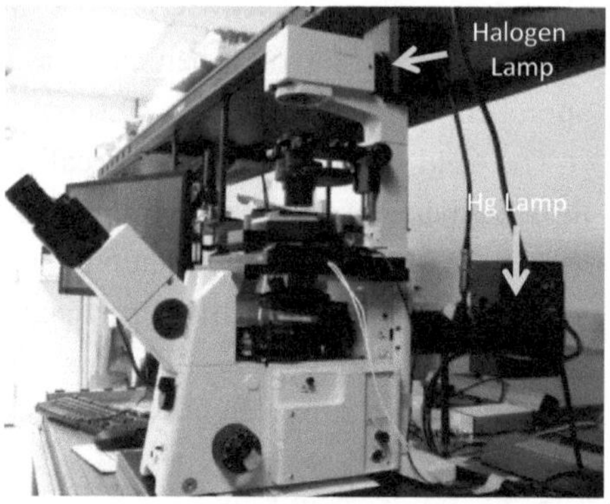

Figure 4.4

Optical microscope with two light sources. One halogen lamp is positioned on the top of the microscope for transmitted white light techniques and one mercury (Hg) lamp is positioned on the bottom of the microscope for epi-fluorescence techniques.

knob on the condenser assembly. Next, center the image of the edges of the field aperture with the centering knobs. The centering knobs translate the condenser lens in the x and y directions to ensure that it is vertically aligned with the microscope objective. The center knobs actually move the lens diagonally, which

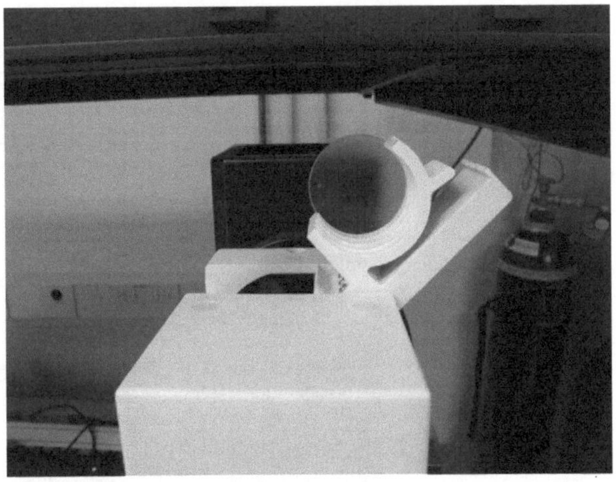

Figure 4.5

A ground glass diffuser is typically located after a halogen bulb to spread out the light and give even illumination to your sample. The diffuser prevents the image of the tungsten filament from showing up in the images you collect.

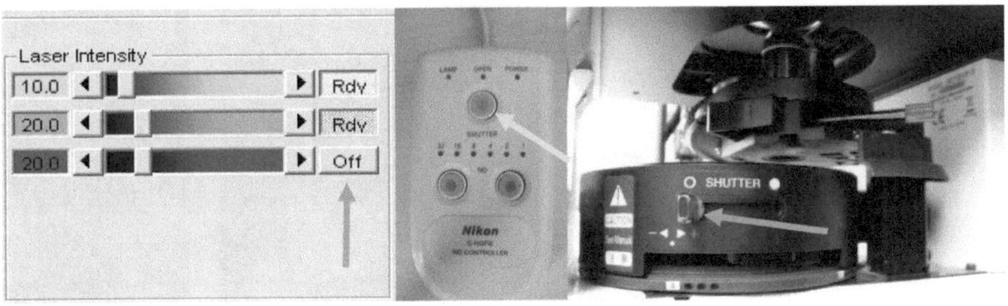

Figure 4.6

A few examples of the locations where shutter controls might hide.

takes a little getting used to. When the image of the field aperture is in focus and aligned, open the field aperture again to just a little bigger than the field of view.

The condenser assembly also includes an adjustable condenser aperture and a turret that can contain phase rings, polarizers, or dark field condensers. We will return to the details of the contents of the condenser assembly turret during the discussion of the relevant techniques throughout this chapter, but for now let us note that it is important to understand the details of the condenser assembly for each imaging modality and continue the tour of the optical path of the microscope. Typically, when starting to set up your experiment, you want to begin in bright field with an empty position of the condenser turret in the optical path.

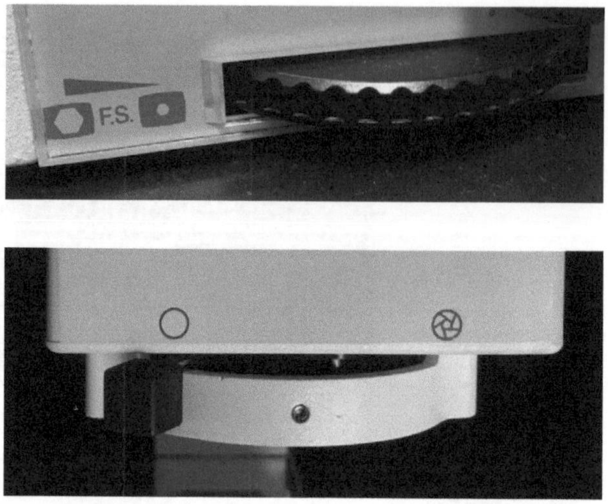

Figure 4.7

Two examples of field aperture (field stop) controls.

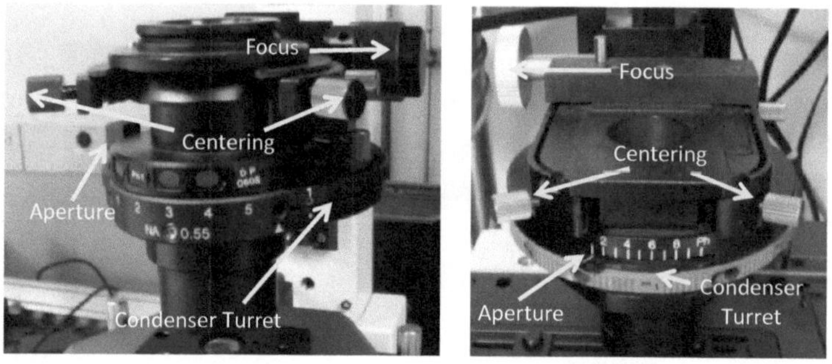

Figure 4.8

Two different microscope condenser assemblies to illustrate similarities and differences between systems.

The sample is the next optical element. If the sample is prepared on a traditional microscope slide, the coverslip should be #1.5. Most microscope objectives are designed to take into account 170 μm of glass before the sample, which is the nominal thickness of a #1.5 coverslip. The coverslip should be facing the microscope objective when the sample is placed on the microscope stage. This means in an inverted microscope the sample must also be inverted. If you are working with a petri dish or a well plate on an inverted microscope, you should use special glass-bottomed dishes or plates with a #1.5 cover glass built in to afford the optimal image quality. Trying to image through a thick slab of polystyrene, especially at high magnification, will be challenging and sometimes impossible.

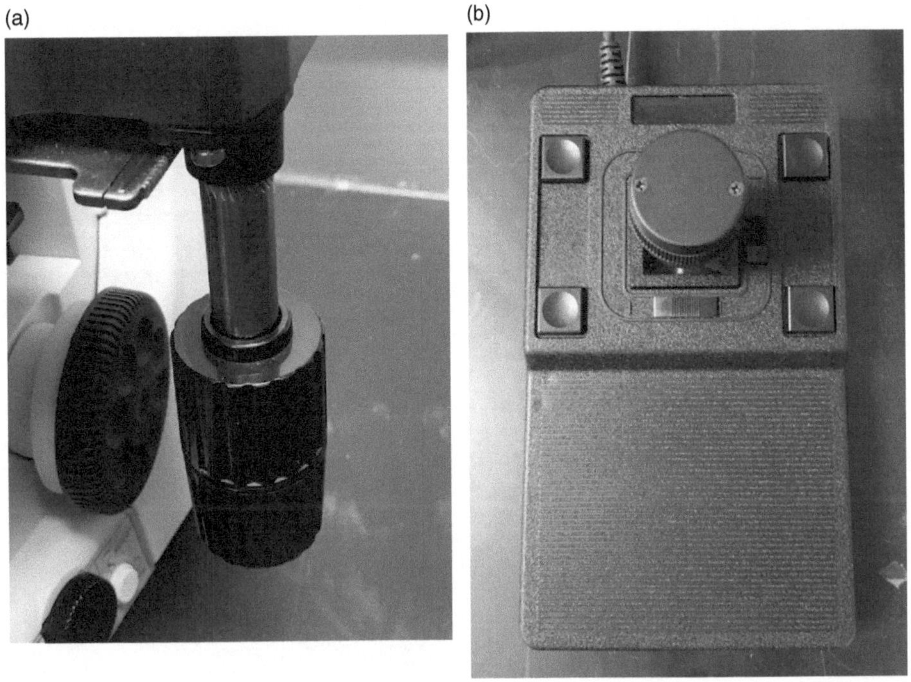

Figure 4.9

(a) Manual stage control. (b) Motorized sample stage control.

If you have a sample preparation that has no coverslip, be sure that you are using appropriate microscope objectives that are designed for there to be no cover glass in the optical path.

The sample sits on the sample stage. Sample stages can be translated manually or they can be motorized. Figuring out how to control the stage before you put your precious sample on it is good practice. Manual stages typically have two knobs hanging vertically off the side to translate the stage in x and y. Motorized stages usually have a joystick.

Finally, we come to the microscope objective (see also Section 2.3.25). Microscope objectives are actually a collection of 10–15 lenses to correct for various optical aberrations. Different objectives are needed for different techniques. Encoded in the markings on the side of the objective are a number of pieces of crucial information. Understanding what microscope objective lenses you have is crucially important to getting an optimal image.

The magnification of the objective is listed numerically on the side of the objective, and is also encoded in a colored ring on the objective. The same color scheme is roughly followed by all microscope vendors (red – 5×; yellow – 10×; green – 20×; light blue – 40×; light blue – 50×; dark blue – 60×; white – 100×).

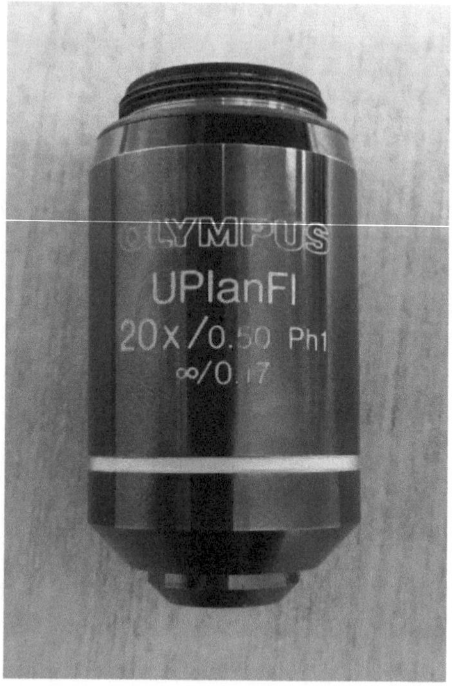

Figure 4.10

An Olympus UPlanFl 20× air objective with a numerical aperture of 0.5. This is an infinity corrected, phase contrast objective designed for cover glass with a thickness of 0.17 mm, with a phase ring 1 in the back focal plane.

The coverslip thickness that each objective was designed for is also listed on the body of the lens, traditionally denoted in millimeters. A typical microscope objective will list 0.17, denoting that it was designed for a 170 μm or #1.5 coverslip to be in the optical path above the sample. Objectives can have a correction collar, which allows you to adjust for different coverslip thicknesses. If the objective that you are using has a correction collar, double-check that it is properly adjusted. There is usually a tiny white tick mark on the stationary part of the objective that can be aligned to different thicknesses of cover glass. (If you take a pair of calipers and measure the thickness of the glass in a box of #1.5 coverslips you will find a bit of variation. To get optimal images you can either measure the coverslips with calipers and only use the 170 μm thick ones, or use a correction collar to compensate for the different thicknesses.)

Microscope objectives also list the numerical aperture (NA) of the objective. The NA is a measure of the solid angle that the objective can collect light from. Higher magnification objectives typically have larger numerical apertures. With an

Figure 4.11

20× objective with a correction collar for different cover glass thicknesses.

air objective, the largest practical NA equals 0.95. (In theory the largest NA for an air objective equals 1, but that would be an impossible perfect lens.) Some lenses use water or air to make an optical bridge with the sample. The immersion fluid will increase the numerical aperture of the objective. Water immersion objectives typically have NAs around 1.3 and oil immersion objectives have NAs around 1.45. The microscope objective lens needs to be designed for each immersion fluid, so please keep track of which objective needs water and which needs oil if you have both on your system.

TIP: Please be sure to clean the oil off the objective whenever you have finished using it. Dried oil on an objective is very hard to remove.

Most microscopes and objectives these days are "infinity corrected." This is denoted on the objective with a small infinity sign (∞). An infinity corrected microscope objective has a parallel light path coming out. Occasionally in labs you will find old objectives that were designed for a specific tube length. These usually won't work well in a modern infinity corrected microscope body.

After the objective there is usually a filter cube turret into which different filter cubes for fluorescence imaging or a polarizer for differential interference contrast (DIC) imaging (Section 4.6) can be located.

Next in the optical path comes a mirror that will direct the light either to a camera or to the eyepieces. Sometimes there will be a filter wheel for fluorescence emission filters directly in front of the camera.

Finally, we come to the detectors. The detector might be your eyes or a point detector or a digital camera. There are pros and cons to each type of detector and different techniques will benefit from different detectors or cameras. Many microscope bodies have a knob that rotates a mirror such that only one detector is receiving light at a time.

4.1.1 Köhler Illumination

There are several microscopy techniques in which the light is transmitted through the sample. The most common of these are bright field, dark field, phase contrast, cross polarized, and DIC microscopies. (Some of these techniques can also be performed in reflection mode on opaque samples.) For any of these transmitted light techniques to work, they need bright, uniform illumination of the sample. The proper procedure for ensuring bright, uniform illumination is known as Köhler illumination. You need to perform the steps to set up Köhler illumination every time you start an experiment and every time you change microscope objectives. The point of the procedure is to ensure that the objective lens and the condenser lens are aligned vertically. If they are offset, then you will be unable to achieve optimal image quality.

The steps to set up Köhler illumination are as follows:

1 Focus on the sample.
2 Close down the field aperture. See Figure 4.7.
3 Focus the condenser so that the image of the field aperture is sharp and clear. See Figure 4.8.
4 Move the image of the field aperture into the center of the field of view by adjusting the position of the condenser lens with two adjustment screws on the condenser assembly that usually move the condenser lens diagonally in the field of view. See Figure 4.8.
5 Open the field aperture to illuminate just to the edges of the full field of view. Do not over-open the field aperture. See Figure 4.7.
6 Look at the back focal plane of the microscope objective. The back focal plane is usually designed to be the back of the microscope objective. There are a number of ways to view the back focal plane. Check to see if the microscope is

equipped with a Bertrand prism or centering telescope. A Bertrand prism is usually inserted into the optical path with a selection wheel on the side of the microscope. Bertrand prisms are usually labeled "B" and should have some sort of focusing slider to bring the image of the back focal plane into focus. A centering telescope is usually located in the turret just below the eyepieces. It is usually labeled "CT" and should also have some sort of focusing slider. If the system you are working with has neither of these components, you can remove the eyepiece (the eyepieces usually just slip out with a gentle pull) and look down the barrel of the microscope to see the back focal plane. This trick works, but is hard as you lose the magnification of the back focal plane gained from the eyepiece.

7 Adjust the condenser aperture to barely fill the back focal plane of the objective.
8 Return to looking at the front focal plane of the objective by moving the Bertrand prism or centering telescope out of the optical path or reinserting the eyepiece.

With a bit of practice, this procedure can be done quickly and easily. Again, if you change objectives (e.g., decide you want to switch from a 20× to a 40× objective), you need to run through all these steps again to ensure optimal alignment for best imaging conditions.

4.1.2 Controlling a Microscope System

As automation on a microscope system increases, so does the complexity of controlling everything. A microscope system purchased as a complete set from a major microscope manufacturer will have its own set of control software to talk to the cameras, motorized stages, motorized filter wheels, etc.

If you are homebuilding a microscope system, there are several third-party options as well. MetaMorph from Molecular Devices is a very powerful third-party system integration software package. The user interface is fairly user-friendly and has good technical support. Unfortunately, MetaMorph is rather expensive. If you are building a system on a budget, Micro-Manager from Open Imaging is a free open-source system integration software package built on top of ImageJ (https://micro-manager.org/). MatLab and LabView could be used too, and many universities already have site licenses for these software packages, but controlling a microscope system with them would certainly take more programming on your part. At the most fundamental level, cheap microcontrollers (Arduino or Raspberry Pi [<$50]) and/or basic scripting languages (e.g., Python, which is free) can also be put to use automating your optical setup's hardware.

4.1.3 Capturing an Image

To collect the best imaging data, check that you are using the full dynamic range of the detector. **Dynamic range** is the number of grayscale values the detector can discriminate between.

A digital camera is a two-dimensional array of photodetectors. Each photodetector is called a pixel. When a photon hits a photodetector, it excites an electron. You can think of each pixel as a small storage bucket for excited electrons. Different cameras have different size storage buckets, both in physical dimensions and in terms of the number of excited electrons that can be stored in each bucket. The number of excited electrons that can be stored is referred to as the bit depth of the camera (Figure 4.12).

The human eye can see ~200 shades of gray. This means that to the human eye an 8-bit image that has 256 grayscale values is fine. Many detectors, however, have a 12-bit, 14-bit, or even 16-bit bit depth. A 16-bit image has 65,536 grayscale values. Finer gradations in grayscale values often allow you to use image processing programs to quantify features of interest more accurately.

Somewhere in the control software for the digital camera, there should be a histogram showing the grayscale values on the x-axis and the number of pixels with that grayscale value on the y-axis. We recommend that the brightest pixel in the image should be about 90 percent of the maximum grayscale value. If a pixel has the maximum grayscale value, it is said to be saturated and the image is overexposed. Keeping the brightest pixel around 90 percent of the maximum gives you a little room to ensure no pixels accidentally saturate. If the brightest pixel is at only 20 percent of the maximum grayscale value, you are underutilizing the potential of your detector and your image is said to be underexposed.

The two parameters that you can adjust to spread the histogram and make sure that you are using the full dynamic range of the detector are the exposure time and the intensity of the excitation light. Increasing both exposure time and/or the intensity of the excitation light will increase the grayscale value of the pixels, and move the histogram toward saturation of the detector.

To save your images, you should always use the TIFF format. TIFF is an uncompressed data format. JPEG is a compressed image format. JPEG images are smaller files because they throw away data that you, as a human who can only see 200 shades of gray, will not notice is missing. You just worked hard preparing your sample, setting up the microscope, and collecting the image, so *please, please,* don't throw away a lot of the data at the last step to save a little bit of computer memory. When you begin image analysis, the additional data contained in the uncompressed format will be useful.

70	69	68	68	70	72	68	63	67	68	70	72	73	73	72	71	74	72	72	74	75
70	70	69	68	69	70	68	64	67	68	69	71	72	73	73	73	76	74	75	78	80
68	69	68	66	66	68	68	66	68	69	69	71	72	74	75	76	78	76	77	80	82
64	66	66	64	63	65	67	67	69	69	70	71	72	74	76	77	79	77	78	81	82
67	65	66	68	68	65	65	66	65	68	73	77	76	70	61	54	71	76	85	89	86
72	69	67	68	68	67	67	68	75	75	75	77	80	82	83	83	88	93	102	107	106
74	70	67	69	71	71	71	73	74	76	80	85	91	97	102	105	114	121	132	140	142
70	67	66	70	74	76	77	78	75	78	85	93	103	114	126	133	151	159	171	179	182
68	66	67	72	77	78	79	80	83	86	93	103	119	140	163	179	188	196	205	210	210
71	70	72	75	77	78	80	83	84	95	113	132	152	174	196	210	207	213	219	222	223
75	75	75	77	77	79	85	92	94	114	143	169	188	201	212	219	217	220	222	222	223
77	77	77	77	78	81	91	101	119	139	167	190	205	213	221	226	229	229	225	220	219
79	82	78	76	83	90	100	112	141	164	192	210	218	221	222	222	226	224	224	226	225
77	78	80	87	96	103	120	142	165	184	207	221	226	228	228	227	224	223	223	225	225
81	82	87	99	110	117	138	169	193	208	224	232	233	232	232	231	226	224	224	226	225
87	93	100	111	126	138	162	189	211	221	231	234	234	233	233	232	231	229	228	229	228
93	108	119	131	152	173	194	211	219	226	232	234	234	235	236	235	236	234	233	232	230
101	125	145	160	181	200	212	219	222	227	232	234	236	239	241	240	239	237	235	235	232
111	140	170	193	208	213	214	218	222	226	230	233	237	241	243	242	241	239	238	238	235
115	146	186	217	229	221	215	222	220	224	228	231	236	241	242	240	243	241	240	240	239
126	155	182	213	226	219	222	226	224	227	231	235	238	240	242	243	243	242	241	239	238
146	176	200	222	230	223	225	226	226	229	233	236	238	240	241	243	243	243	242	241	239
167	199	218	230	233	226	228	227	229	232	235	237	239	240	241	243	244	244	244	244	242
181	214	230	235	235	231	234	231	234	236	238	239	240	241	242	243	245	245	246	246	245

Figure 4.12

A digital image is really an array of intensity values. An 8-bit image has values between 0 and 255. A value of 0 denotes black and a value of 255 denotes white.

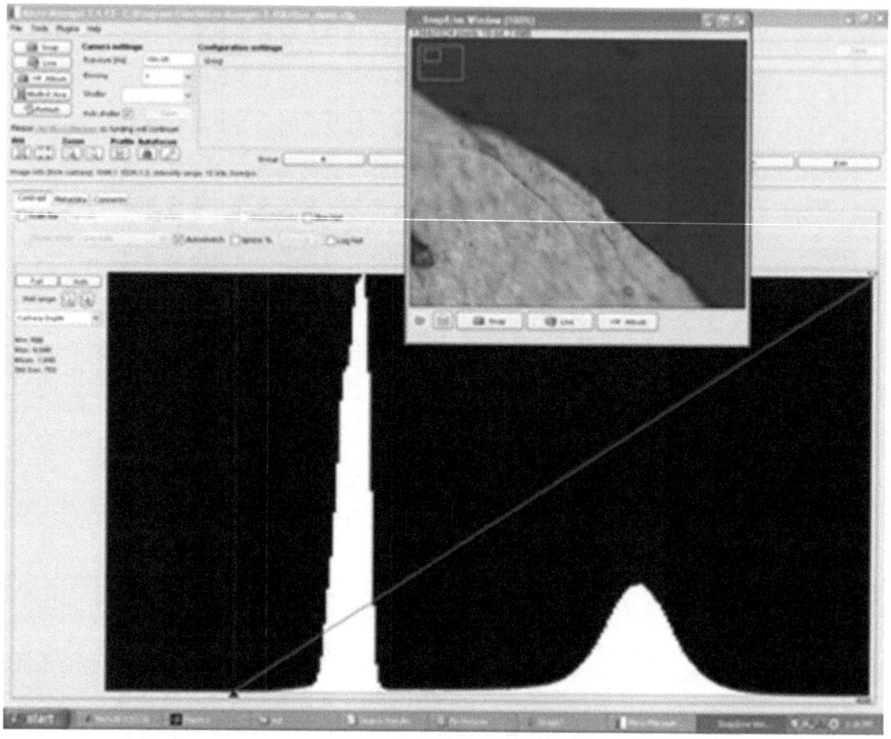

Figure 4.13

A histogram of grayscale values showing a properly exposed image. The *x*-axis is the intensity values and the *y*-axis is the number of pixels with that grayscale value.

4.1.4 Analyzing Images

After you collect images, you may want to quantify some of the information contained in them. There are many different options for processing your data. The control software for microscope systems often has some data-processing features incorporated. There are a number of third party image analysis software packages such as Imaris, Arivis, or Avizo. MatLab and Mathematica also have image processing capabilities. However, many of these commercial programs are expensive and have restrictive licenses. The commercial programs often handle very large data sets better than the open source programs though.

One free, yet quite powerful, image processing program is called ImageJ. ImageJ is an open-source program that was originally sponsored by the National Institutes of Health. The NIH still maintains an official ImageJ website, but the Fiji package of ImageJ is often considered a bit more user-friendly and tends to be better documented. Fiji can be downloaded at Fiji.sc. ImageJ and Fiji both include scripting languages that allow automated image processing. Another free open source image analysis package specifically for biological microscopy is called CellProfiler and can be found at cellprofiler.org.

4.2 Bright Field Imaging

Underlying Physical Principles

Bright field imaging is the simplest imaging modality. In bright field images, the objects of interest appear dark on a bright background field (hence the name bright field). Contrast is generated by the object of interest absorbing light, so the intensity of light at the position of the object is diminished. Bright field microscopy can be performed in transmission or in reflection modes.

If the object of interest does not naturally absorb light, a stain that does absorb light can be attached to it. This is especially important in biology. Histological slides that are hematoxylin and eosin (H&E) stained are a common example of the use of bright field imaging. H&E stained slides imaged in bright field microscopy are still the gold standard technique for imaging in pathology.

Performing Bright Field Imaging

To get the best bright field image, it is absolutely necessary to set up Köhler illumination. This procedure aligns the condenser lens with the microscope objective along the optical path. See Section 4.1.1 for the steps to set up Köhler illumination.

Strengths and Limitations

The biggest limitation of bright field imaging is the requirement that the structures of interest absorb light. Many objects of interest are transparent, and these will show up with limited contrast in bright field images.

What Samples Are Appropriate?

Any sample that has some color, high absorption, natural contrast, or is stained will work well for bright field. For bright field in transmission mode, clear samples are very hard to image. Samples should be mounted on a typical microscope slide. In reflection mode, samples can be anything. Make sure that the microscope objective that you are using is designed to work either with or without a coverslip depending on how you have mounted your sample.

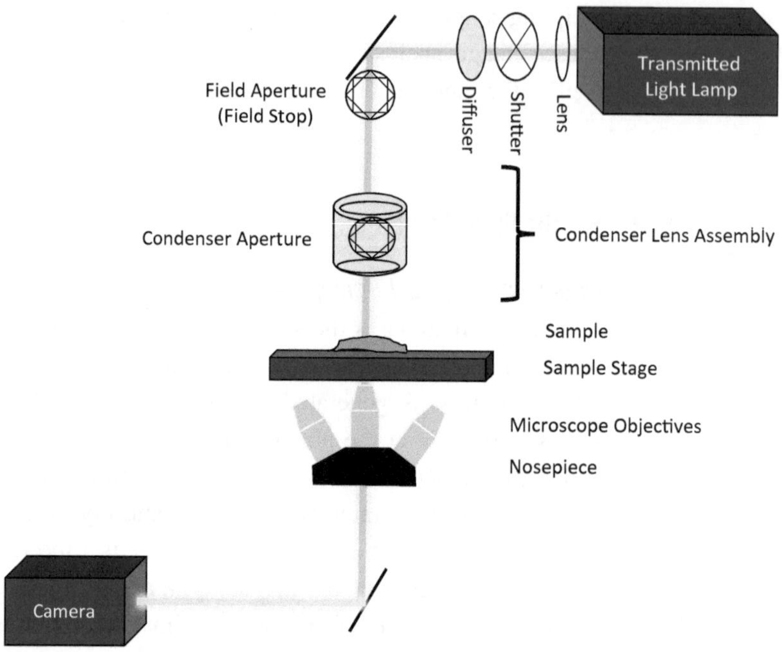

Figure 4.14

Layout of transmitted bright field microscope setup.

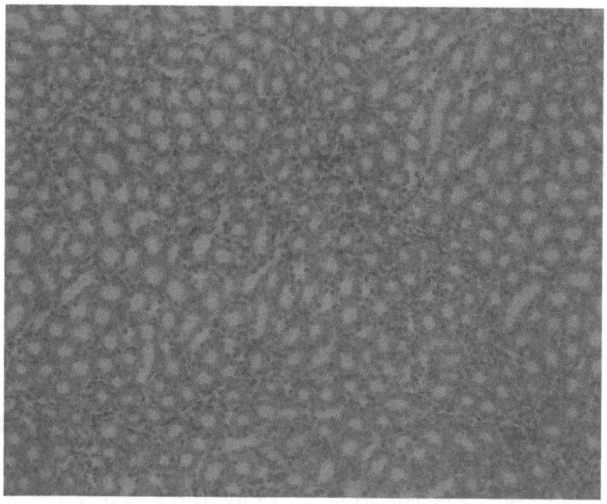

Figure 4.15

20× magnification of stained rat kidney sample in bright field transmitted light.

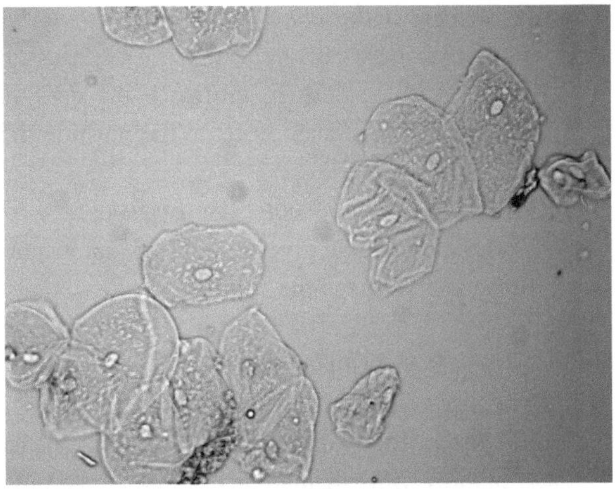

Figure 4.16

The author's cheek cells imaged with bright field microscopy. Unstained cells and tissue have low contrast and are challenging to see in bright field.

4.3 Dark Field Imaging

One challenge with bright field imaging is that the contrast, the difference in signal intensity between the areas of interest and the background, can be low. Dark field microscopy improves contrast by throwing away all the photons that did not interact strongly with your sample. This leads to a very low black background in the image. The structures of interest in the sample need to scatter light strongly though. Dark field microscopy can be performed in transmission or in reflection modes.

Underlying Physical Principles

In dark field images, the objects of interest appear bright on a dark background field (hence the name dark field). In dark field microscopy, a dark field stop is introduced into the condenser to block the light directly along the optical axis. This illuminates the sample with a hollow cone of light. This oblique angle illumination of the sample is done so that the light that is simply transmitted through the sample is not collected by the microscope objective. These transmitted photons often contain little information about the sample anyway, so throwing them away can improve contrast. Only photons that are strongly scattered by the sample back into the optical axis will be collected by the microscope objective. These are the photons that interacted strongly with the sample and thus contain more information about the sample.

Performing Dark Field Imaging

To get the best dark field image it is necessary to set up Köhler illumination in bright field first. Then, introduce the dark field stop into the optical path by rotating the condenser turret to the position that contains the dark field stop.

What Scientific Questions Can Be Asked?

Dark field imaging gives you the spatial location of highly scattering structures or objects in your sample.

Strengths and Limitations

The structures of interest in the sample must be highly scattering for dark field microscopy to work. For this reason, one can often localize sub-wavelength structures while maintaining high contrast in the image.

What Samples Are Appropriate?

The samples must be highly scattering for dark field to work well. Metal nano-particles in a biological tissue is a classic example of a sample that works well in dark field. Most inorganic materials with surface roughness and/or high surface area to volume ratios (e.g., nanoscale structures) will provide good scattering

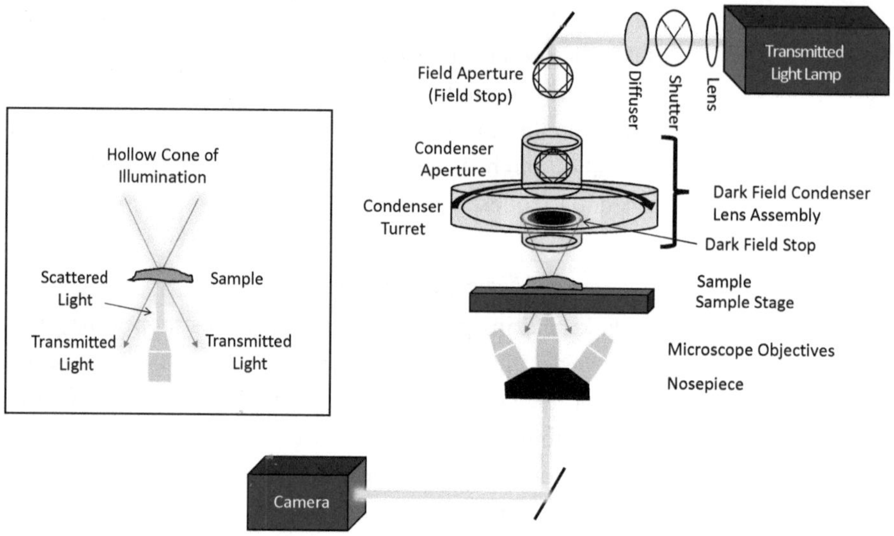

Figure 4.17

Dark field microscopy setup. In dark field imaging, a dark field stop is rotated or translated into the optical path to illuminate the sample at a highly oblique angle. The light that is transmitted through the sample is intentionally not captured by the microscope objective. The microscope objective only captures the light that is highly scattered by the sample.

Figure 4.18

Dark field condenser stop.

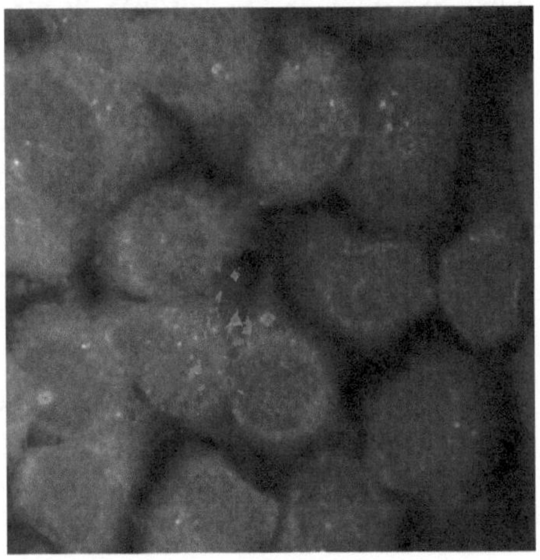

Figure 4.19

Darkfield image of gold nanoparticles in HeLa cells. The small bright dots are the gold nanoparticles.

contrast. In transmission mode, samples should be mounted on a standard microscope slide. Again, be sure that you have selected appropriate microscope objectives for your sample mounting. Is the microscope objective designed to work with a coverslip or not?

4.4 Phase Contrast Imaging

Phase contrast imaging is a method of generating contrast based on the phase change of light as it passes through a sample. This technique is a transmitted light technique that is typically applied to unstained, transparent samples. It requires a phase annulus to be placed in the condenser and a special microscope objective with a phase ring in the back focal plane. Phase contrast can only be done in transmission mode.

Phase contrast is commonly used in **tissue culture microscopes**. A tissue culture microscope is a microscope used to check on cell growth in flasks, well plates, or petri dishes. Tissue culture microscopes and phase contrast imaging are designed to work with the thick plastic walls of a T75 flask. The alignment of the phase contrast optical components is usually simplified, but you should still take a moment to identify the optical components before using the tissue culture microscope.

Underlying Physical Principles

Photodetectors are not sensitive to the phase of the electric field in the electromagnetic wave known as light. Rather, photodetectors are only sensitive to the amplitude squared of the electric field (also known as the intensity of the light). For transparent samples (such as cultured cells), there is little intensity change when the light passes through the sample. But, there is often a phase change though as the index of refraction of the sample is slightly different than that of the surroundings. Phase contrast imaging uses some extra optical elements to convert differences in phase to differences in intensity that can be detected by the digital camera or the human eye.

To convert differences in the phase of light to differences in the intensity of light, several optical elements must be added to the optical path. First is the condenser annulus. This is a mask with a donut shape cut into it which is inserted in the excitation light path before the sample (Figure 4.23). The point of the condenser annulus is to only illuminate the sample with a ring of light. The ring of light is set up so that it will still be collected by the microscope objective, which is a key difference between dark field and phase contrast imaging. The condenser annulus is located in the condenser turret so it can be easily rotated in and out of the light path. There are usually several differently sized annuli in the condenser turret for phase contrast microscopes. They will be labeled Ph 0, Ph 1, etc. This slightly cryptic labeling denotes the different standard sizes of the annuli.

On the collection light path after the sample, you will need to use a special phase contrast microscope objective. The phase contrast objective will be labeled

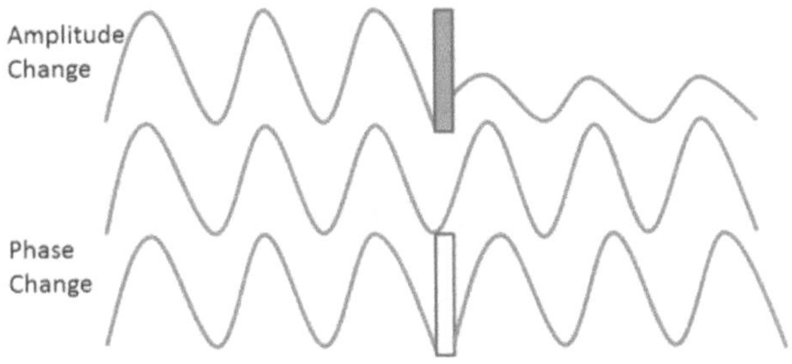

Figure 4.20

Top: Intensity changes caused by light being absorbed by the sample can be detected by cameras and human eyes. Bottom: Phase changes caused by transparent samples are not detectable by cameras or the human eye.

Ph 0, Ph 1, etc. to denote the size of the condenser annulus that needs to be used with it. The backs of phase contrast objectives include a dark ring. The size of the dark ring will need to exactly match the size of the condenser annulus

The purpose of the dark ring is to attenuate the light that had no interaction with the sample and simply passed straight through the sample. The light that didn't interact with the sample contains no information about the sample and therefore only contributes to the background in the image, decreasing contrast. The phase ring also introduces a $\lambda/4$ phase delay to the transmitted light. This usually puts it $\lambda/2$ out of phase with the scattered light, giving maximum destructive interference.

While the hardware can look confusingly similar between dark field and phase contrast microscopy, the key difference between the two techniques is that in dark field microscopy, the transmitted light is excluded from the optical collection path and only the scattered light is observed. In phase contrast microscopy, the transmitted light is collected, attenuated, and interfered with the scattered light.

Performing Phase Contrast Imaging

To acquire the best phase contrast image, first set up Köhler illumination in bright field using a phase contrast objective. Then, introduce the condenser annulus to the optical path by rotating the condenser turret to the proper position.

Finally, you will need to check the alignment of the phase ring and the condenser annulus. To check the alignment, look at the back focal plane of the microscope objective. There are several ways to do this. One method uses a centering telescope, which is a set of optics mounted in a wheel immediately behind the eyepiece, usually labeled CT. Rotate the centering telescope into position and adjust the focus to get a clear image of the dark ring on the back of the objective and the bright ring of illumination from the condenser annulus. A Bertrand prism provides another way to view the back focal plane. The Bertrand prism is usually mounted in a wheel on the

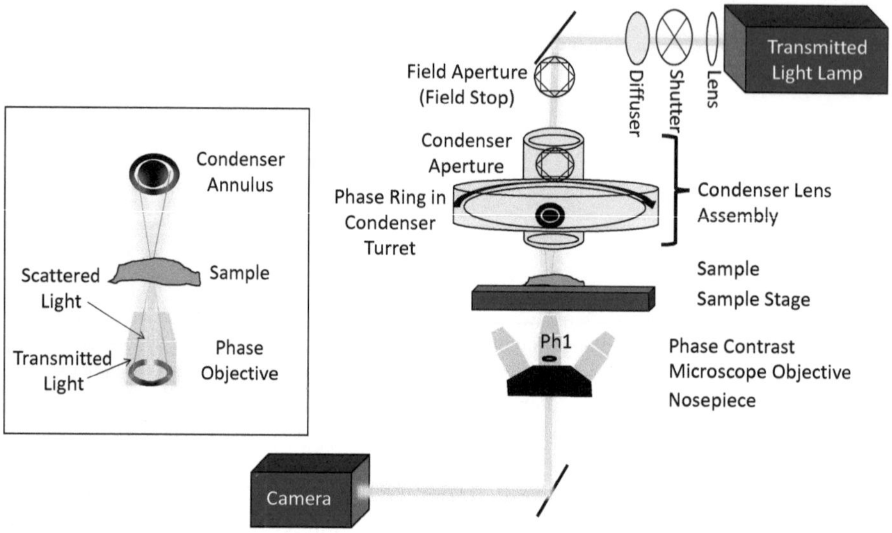

Figure 4.21

Schematic of a phase contrast microscope setup. The transmitted light that passes straight through the sample without interacting is collected by the objective but is strongly attenuated by the phase ring in the microscope objective. It then interferes with the light that was scattered by the sample to convert the phase shift caused by the sample into an intensity change that can be detected by the camera.

side of the microscope labeled B. Rotate the Bertrand prism into position and again focus to generate a clear image of the dark ring on the back of the objective and the bright ring of illumination from the condenser annulus. These two extra sets of optics are advantageous because they enable a magnified image of the back focal plane of the objective.

If the microscope that you are working with does not have either a center telescope or a Bertrand prism, you can simply remove one of the eyepieces of the microscope and look down the barrel of the microscope. (Typically microscope eyepieces slip out with a gentle tug.) You will see a small dark ring and a bright ring. To align the bright ring to the dark ring, adjust the position of the condenser annulus in the condenser turret. There will be some small unmarked screws that will move the condenser annulus diagonally in the field of view. Sometimes they will be set screws that you will need to use a hex wrench to adjust. The bright ring and the dark ring should overlap exactly. When you are done aligning the phase ring and the condenser annulus, rotate the center telescope or Bertrand prism out of position or reinsert the eyepiece. Your image should have the signature phase contrast halo around the structures. If it doesn't, the setup is configured incorrectly. Please work your way through the setup procedure again.

Figure 4.22

Phase contrast objective with a Ph 1 phase ring.

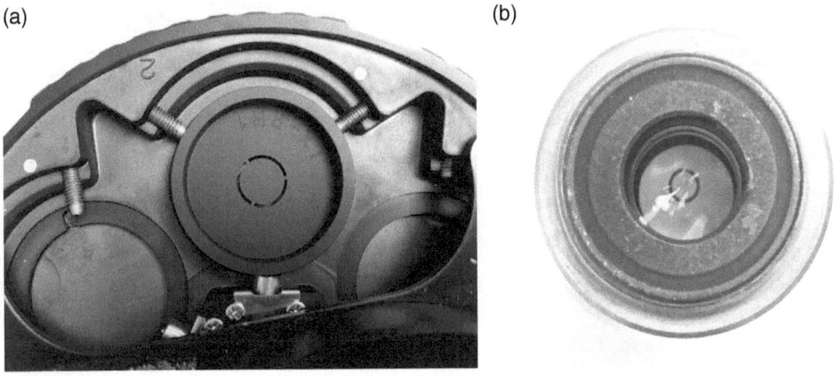

Figure 4.23

(a) Condenser annulus Ph 1. (b) Phase ring in back focal plane of microscope objective.

What Scientific Questions Can Be Asked?

Phase contrast gives the spatial location of phase objects – that is structures or objects in the sample that change the phase of light as it is transmitted through the sample. Phase contrast imaging is predominately used in biology for unlabeled cultured cells. In materials science, users can also characterize transparent samples such as polymers and organic electronics.

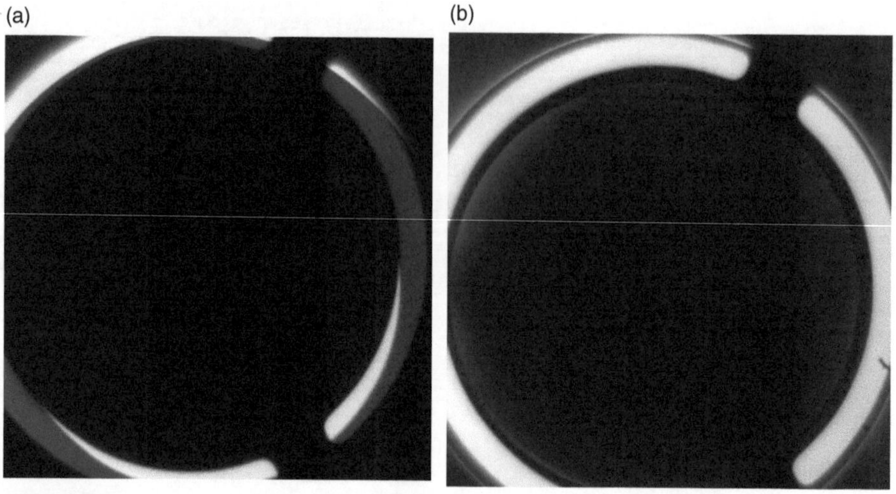

Figure 4.24

(a) Misaligned phase ring and condenser annulus. (b) Properly aligned phase ring and condenser annulus.

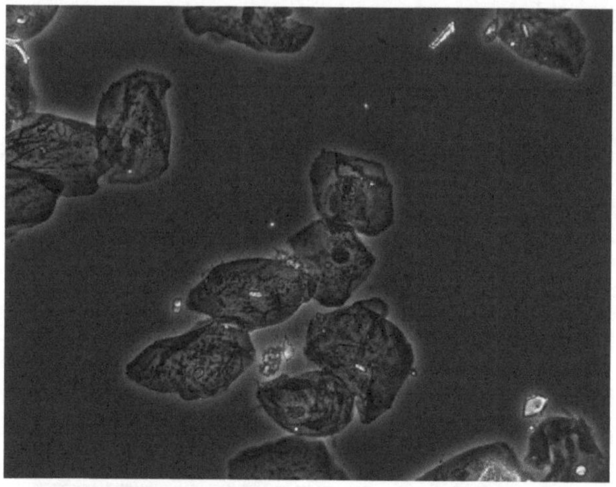

Figure 4.25

Cheek cells under phase contrast. Note the characteristic halo around the edge of the cells. If phase contrast is properly set up, your images should have this halo. If there is no halo, phase contrast imaging is not configured correctly.

Strengths and Limitations

There is a signature halo around the objects when phase contrast is properly executed. This can be a limitation depending on exactly what your scientific question should be.

Common Pitfalls

Be sure the phase ring and the condenser annulus are properly aligned, per the above steps.

4.5 Cross Polarized Imaging

Cross polarized imaging takes advantage of the sample changing the polarization of light as light interacts with the sample. This mechanism produces extra contrast for birefringent structures in your sample. Cross polarized microscopy can be performed in transmission or in reflection modes.

Underlying Physical Principles

The birefringence of a sample can be used as a contrast mechanism in cross polarized imaging. In cross polarized imaging, a linear polarizer is introduced to the excitation light path. Another linear polarizer (referred to as the analyzer) is introduced to the collection light path and set at 90 degrees to the excitation polarizer. If the sample has no birefringence, the polarization of light will not change as it propagates through the sample and the analyzer will block all the light. In this scenario, you will simply record a black image. However, if the sample has some birefringence, the polarization of the light will change after it interacts with the sample and the light with changed polarization will be transmitted through the analyzer to the detector. Cross polarized imaging can be configured in transmission or in reflection modes.

Performing Cross Polarized Imaging

To get the best crossed polarized image, it is necessary to set up Köhler illumination in bright field first with a strain-free objective. Next, introduce the polarizer and analyzer into the optical path. Finally, cross the polarizer and analyzer to make the background of the image black. If you cannot get the background to go black, you probably have some sort of sample mounting issue or you are using an inappropriate objective which has some birefringence itself.

What Scientific Questions Can Be Asked?

Cross polarized imaging reveals the nature of the birefringence of the sample, which can be informative about crystal structure. This technique is widely used in petrology (the study of rocks) and crystallography for this reason. Specific biological structures also exhibit birefringence and can be studied with cross polarized imaging. Strain in plastics causes birefringence and can be imaged in cross polarized microscopy (see Figure 1.17).

Strengths and Limitations

Cross polarized microscopes give high-contrast images. Structures of interest in the samples must be birefringent to be seen in a cross polarized microscope.

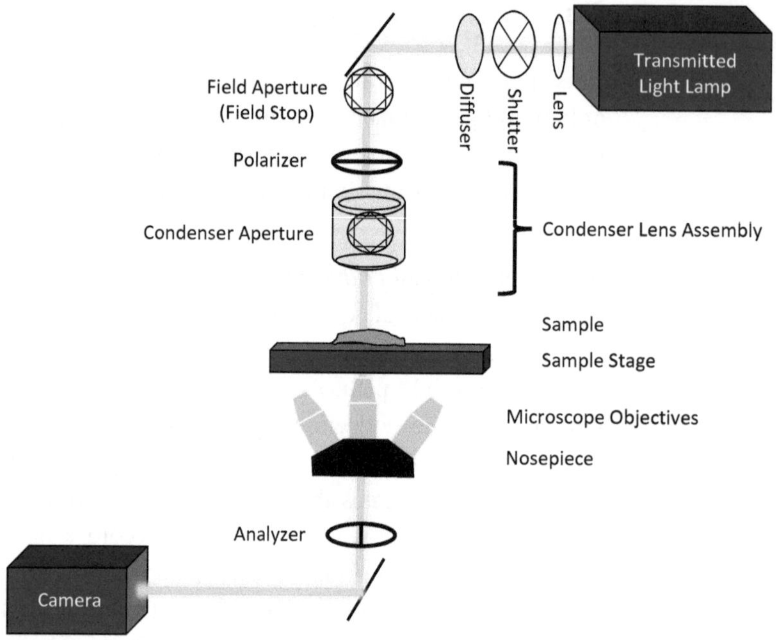

Figure 4.26

Transmission cross polarized microscope setup.

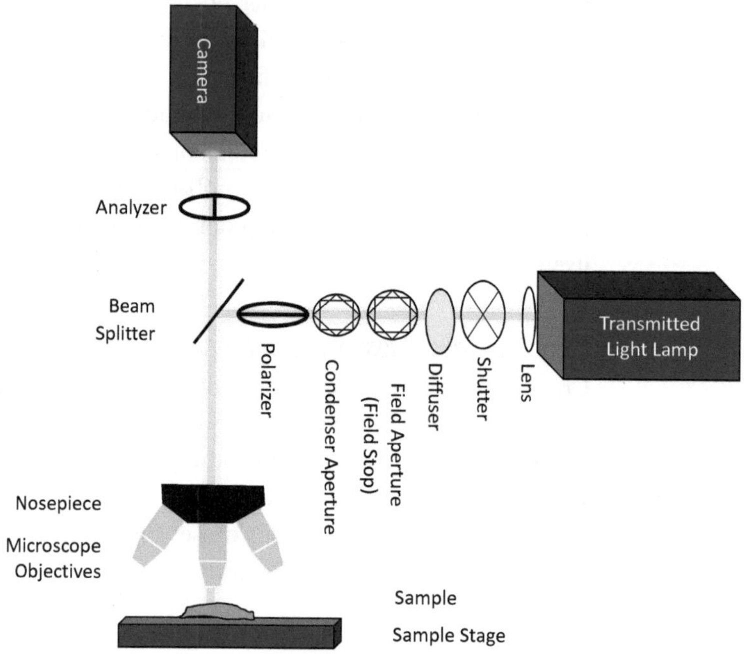

Figure 4.27

Cross polarized microscope setup in reflection mode.

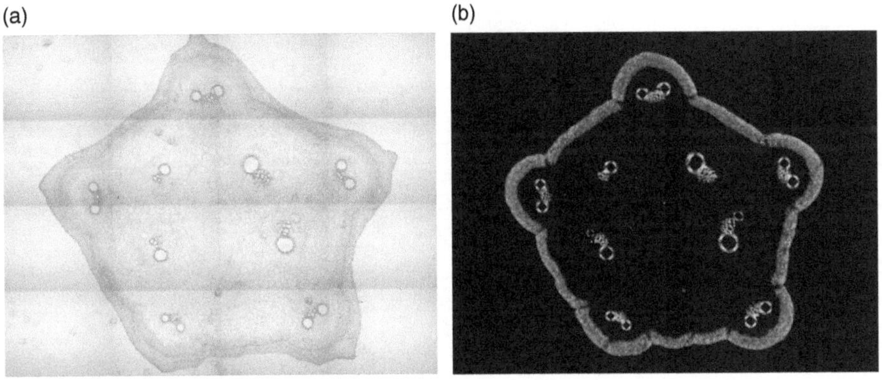

Figure 4.28

(a) Bright field image of stained pumpkin stem. (b) Cross polarized image of pumpkin stem, highlighting structures in the plant with birefringence.

Common Pitfalls

Not being careful about your sample preparation can cause problems in cross polarized microscopy. For example, if a birefringent mounting medium is accidentally used to mount the sample you will not get a meaningful image of your sample. Using a plastic petri dish or a plastic slide or plastic coverslip will cause problems.

Microscope objectives that are not strain-free will introduce their own birefringence into the image. You must use objectives that are manufactured to have strain-free lenses. They are usually denoted DIC or POL on the objective.

4.6 Differential Interference Contrast Imaging

Differential interference contrast (DIC) imaging is one of the most popular transmitted light techniques, as it gives a false sense of three dimensionality to the image that provides broad aesthetic appeal. Note that DIC is nontrivial to set up properly, and is often configured incorrectly. DIC microscopy can be performed in transmission or in reflection modes.

Underlying Physical Principles

In DIC imaging, the incoming excitation light is polarized. A birefringent prism (either a Nomarski or Wollaston prism) then splits the excitation light into two spatially separated beams, known as the ordinary (o) and extraordinary (e) beams. The two beams then interact with the sample. After the sample interaction, the two beams are recombined into a single beam with another birefringent prism. The

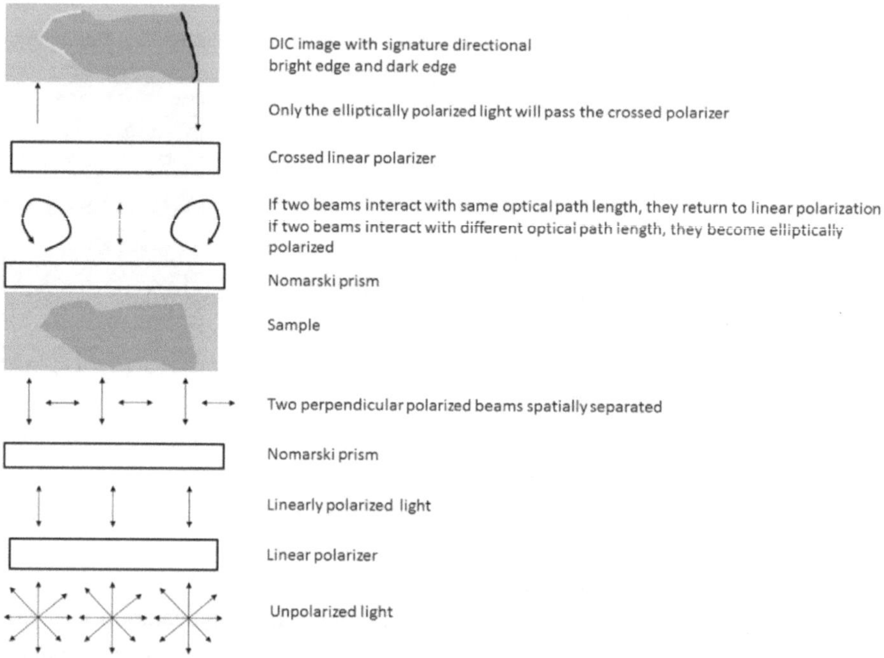

Figure 4.29

Origin of DIC contrast.

single recombined beam is then sent through a second polarizer (referred to as the analyzer since it is in the collection light path) that is crossed with the first polarizer in the excitation path. Finally, the light is sent to a detector.

If the two beams interacted with exactly the same optical path in the sample, they will be perfectly recombined by the second prism and blocked by the analyzer. If the two beams interacted with different optical paths (like at an edge in the sample), they will be recombined into an elliptically polarized beam instead of a linearly polarized beam, and some portion of the light will be transmitted through the analyzer.

Differential interference contrast microscopy is sometimes call Nomarski microscopy. It can be performed either in transmission or in reflection geometries.

Performing Differential Interference Contrast Imaging

To set up DIC imaging, first focus on the sample and set-up Köhler illumination in bright field configuration. Next, introduce the polarizer and analyzer and cross them. The image should be completely dark when they are cross polarized. If the image is not completely dark under cross polarization, you have birefringence in your sample and it is not suitable for DIC. Introduce the two birefringent prisms. For transmission DIC configurations, one birefringent prism is often hiding in the condenser turret and one is hiding right behind the microscope objective. For reflection DIC configurations, one birefringent prism serves double duty and is usually located just behind the objective lens. In both transmission and reflection DIC configurations the birefringent prism that is right behind the objective usually has a small knob on it that allows you to adjust the amount of the prism in the beam path. By adjusting the amount of the prism in the beam path, you change the spatial separation between the ordinary and extraordinary beams. This spatial separation is called the "shear." Adjust the shear to get the best contrast on the features of interest in your sample. When everything is adjusted correctly, you should have a bright edge and a dark edge to each feature on a gray background. If you do not have a bright edge and a dark edge to each feature, something is not adjusted correctly. Please start over.

What Scientific Questions Can Be Asked?

DIC provides contrast at the edges of optically transparent samples. The shadow cast in the image is not real three-dimensional information, but it does provide a hint at the sample's topography, and an opportunity to acquire high-contrast

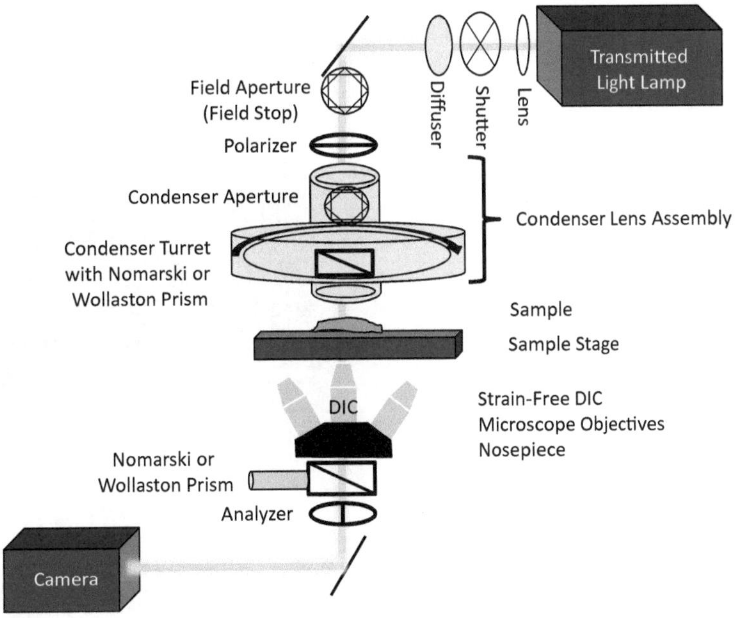

Figure 4.30

Transmitted DIC microscope configuration.

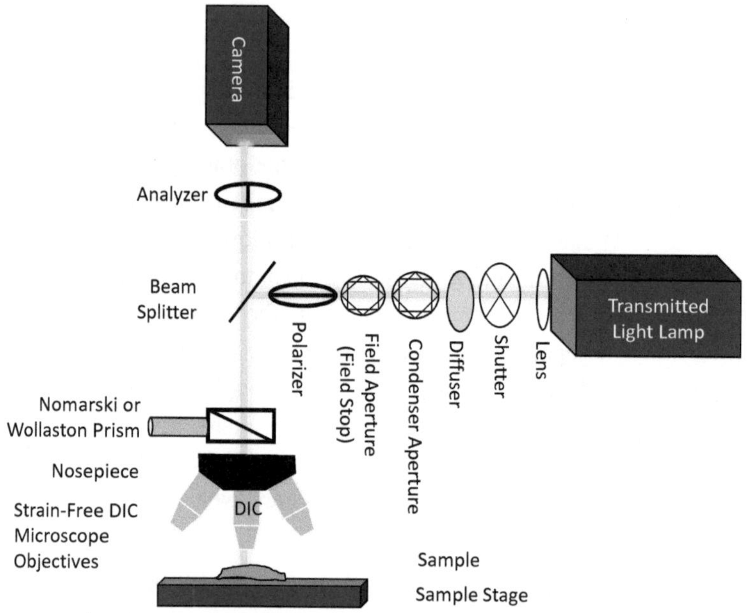

Figure 4.31

Reflected DIC microscope configuration. The Nomarski prism does double duty here by both splitting and recombining the beam.

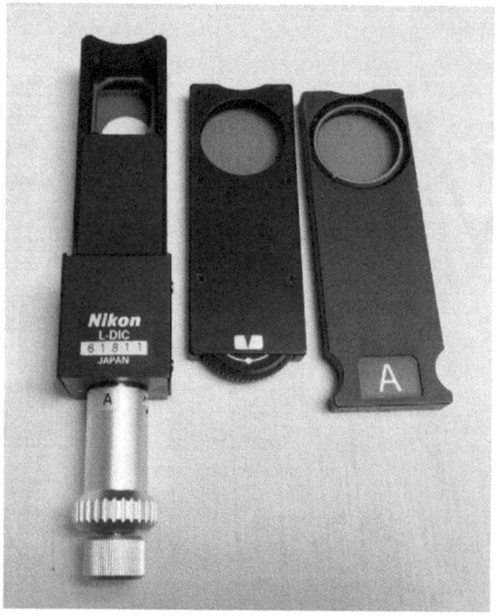

Figure 4.32

Left, DIC prism; center, polarizer; right, analyzer.

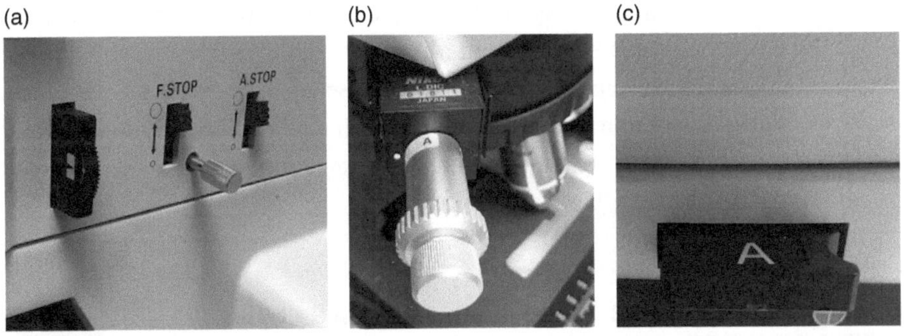

Figure 4.33

(a) Polarizer in an excitation path. The notched wheel allows the polarizer to be rotated to set up a cross polarized configuration. (b) DIC prism in place directly behind the objective. (c) Analyzer in place in the collection path.

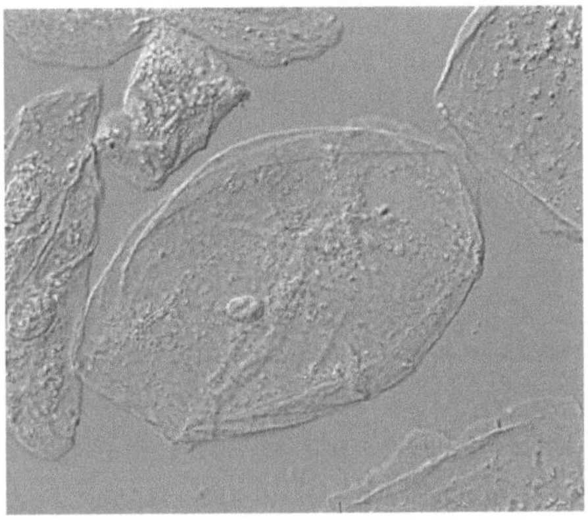

Figure 4.34

Differential interference contrast image of cheek cells from one of the authors. Note the bright and dark edges that are a signature of properly executed DIC imaging.

visualization of samples that might be very difficult to image with other techniques. DIC can be used for transparent biological or transparent materials samples.

What Samples Are Appropriate?

Samples must be transparent with no birefringence. Samples for transmission DIC should be mounted on glass slides or placed in a glass-bottomed dish. Plastic petri dishes or plastic well plates will not work for transmission DIC.

Common Pitfalls

Differential interference contrast imaging is time-consuming to set up. Rushing through the process will likely result in a less than optimal image. If there is any birefringence in the sample or in the objective, you will not get a DIC image. If you are working in a dish with cells, make sure you have a glass-bottomed dish. For the same reason, do not use a plastic lid on the petri dish. Instead, float a layer of mineral oil on top of the cell media to keep samples from drying out. Check that there are the signature bright and dark edges to the features to ensure that you set everything up correctly.

4.7 Wide Field Fluorescence Imaging

Wide field fluorescence microscopy is excellent for visualizing flat, fluorescently labeled samples. Fluorescently labeling samples yields high-contrast images since the signal only comes from the structures that are labeled, and the background is dark. Wide field fluorescence microscopy is comparable to wide field photoluminescence imaging (Section 3.3).

Underlying Physical Principles

A fluorescent molecule absorbs light of one color and emits light of a different color, allowing discrimination between excitation light and the fluorescence signal light. As shown in Figure 4.35, the excitation photon can be absorbed by the molecule to move an electron from a ground state to an excited state. The electron will lose some energy due to non-radiative relaxation as it relaxes to the bottom of the excited state energy manifold. The molecule will then emit a new photon of a lower energy as the electron returns to the ground state energy manifold.

Fluorescent molecules can be attached to specific (usually biological) structures to enhance contrast. Since the excitation light is a different color than the emitted light, they can be easily separated with optical filters. This gives a high-contrast image as the background is now almost zero and the fluorescent signals can be made quite bright.

4.7.1 Sample Preparation for Fluorescent Imaging

Fluorescent dyes can be specifically located in a cell by using immunocytochemistry. In essence, immunocytochemistry leverages the molecular specificity of an antibody produced by an immune system to help label biological structures of interest. Primary antibodies are produced by injecting a protein of interest into an animal of a different species. The animal's immune system produces antibodies against that protein as a defense mechanism. These antibodies are then harvested. Fluorescently labeling biological structures is complicated and countless techniques have been developed. For more detail than is provided in this section's brief overview, please refer to the references listed in Section 4.21.

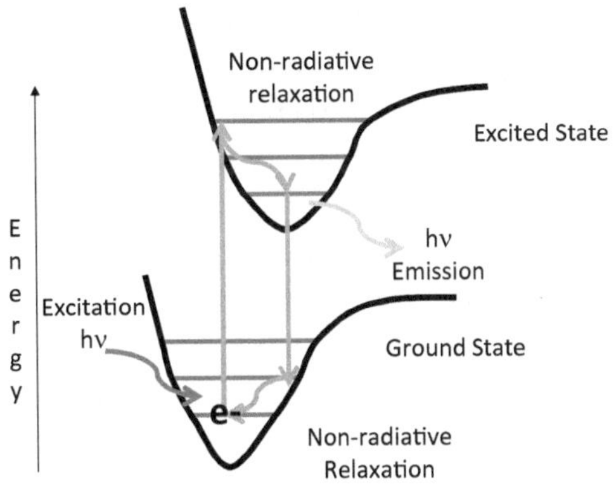

Figure 4.35

Fluorescence cycle of a fluorophore.

Secondary proteins are usually produced by taking the immunoglobulin G (IgG) protein, a portion of the antibody protein, from the animal that produced the primary antibody and injecting it into an animal of a different species. This animal's immune system will then produce antibodies that will attach to the primary antibody. The secondary antibodies are harvested and then chemically attached to the fluorescent dyes. Using secondary antibody labeling allows for more fluorophores to be attached to the structure of interest, providing a brighter image.

For traditionally labeled fixed biological microscope slides, the cells are first "fixed," which means killing the cells and then introducing a cross-linking polymer (usually formaldehyde) that will preserve the cellular structure of the sample. The primary antibodies are then introduced and they attach to the protein of interest and provide a binding site for the secondary antibodies that are fluorescently labeled. The secondary antibodies are then introduced and bind to the primary antibodies. Multiple secondary antibodies can attach to a single primary antibody, providing higher density of fluorescent labeling. Secondary antibodies will bind to any antibody produced by the species of animal that produced the primary antibodies, allowing secondary antibodies to be a little more versatile.

The nomenclature of immunocytochemistry can get a little confusing. If a primary antibody for tubulin was produced by a rabbit's immune system, it would be called a "rabbit anti-tubulin" primary antibody. A secondary antibody produced in a donkey against a generic rabbit antibody and labeled with Alexa 488 fluorescent dye would be termed an "Alexa 488 donkey anti-rabbit" secondary antibody.

Fluorescent dye names often include the excitation wavelength that is most commonly used to excite them. The Alexa dyes are a popular common set of dyes that have been modified to emit throughout most of the visible spectral region. The

488 denotes that a 488 nm laser is typically used to excite the dye. The number 488 nm may seem a little arbitrary, but it is a commonly used laser line of an argon ion gas laser (see Section 2.2.3).

Another common labeling mechanism is the use of fluorescent stains, which are molecules that will specifically bind to a biological structure without an antibody. DAPI, which binds to double stranded DNA, is the most common and widely used stain. Since DNA resides in cell nuclei, DAPI is used to visualize cell nuclei.

For live cell imaging, genetic code can be introduced into the cells that cause fluorescent proteins to be manufactured by the cell and attached to the structure of interest. Fluorescent proteins started with the discovery that the genetic code for green fluorescent protein (GFP) found in a jellyfish in the Pacific Ocean could be inserted into other types of cells with apparently no ill effect. GFP has since been mutated to fluoresce in almost any color desired.

Fluorescent imaging is almost always performed in a standard optical microscope body in epi-illumination (also known as "epi-fluorescence imaging"), which means that the excitation and the signal are collected from the same side of the sample. This configuration allows the residual excitation light that did not interact with the sample to be dumped in the direction away from the detector. A filter cube is placed inside the filter cube turret (Figures 4.37 and 4.38). A filter cube consists of three filters. First is an excitation filter that selects out a single wavelength from a white light source or spectrally "cleans up" the wavelength from a single wavelength source such as an LED or laser. Next is a dichroic filter

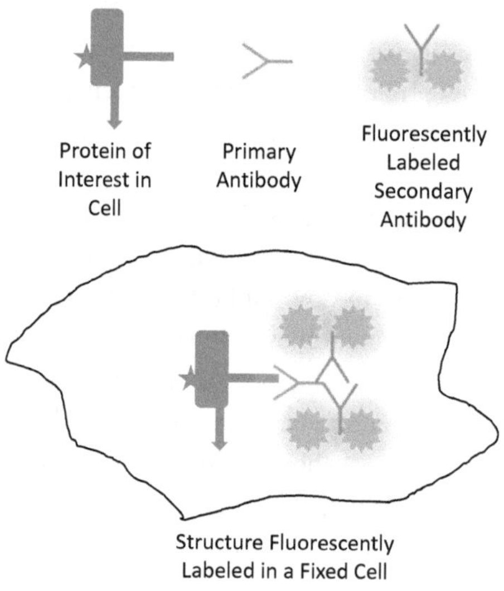

Figure 4.36

Cartoon of immunofluorescently labeled cell.

that will reflect the excitation light to the sample but pass the fluorescent signal light coming back from the sample. Finally, an emission filter will help spectrally "clean up" the fluorescent signal by rejecting the excitation wavelength and passing the emission wavelength.

A typical filter cube turret has six positions, allowing for different filter sets to be moved into the optical path without disrupting the sample position or optical alignment. In some systems the excitation and emission filters are moved out of the filter cube into filter wheels directly after the excitation source and directly before the detector. In this case, the dichroic filter that is left in the filter cube is usually designed for multiple spectral regions with up to four different dyes being accommodated. The advantage of this configuration is that it allows for rapid and easy switching of the excitation and emission wavelengths. You still have to take the multiple images sequentially, changing filters each time, but an external filter wheel is faster than an internal motorized filter turret. Note, however, that with an external filter wheel, the emission filters are no longer in the optical path for the microscope's eyepieces. So, you can only view the fluorescence image on the computer screen through the digital camera.

TIP: In an inverted microscope, a small box placed over your sample on the sample stage will help train the residual excitation light, which can be dangerous if it is a laser source. The box will also help reject room lights, computer monitor light, etc. from being collected by the objective and sent to the detector.

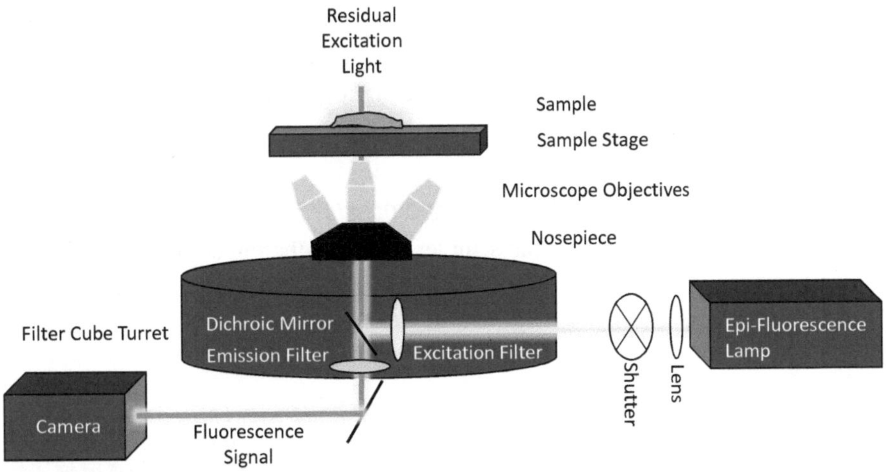

Figure 4.37

Layout of an epi-fluorescence microscope.

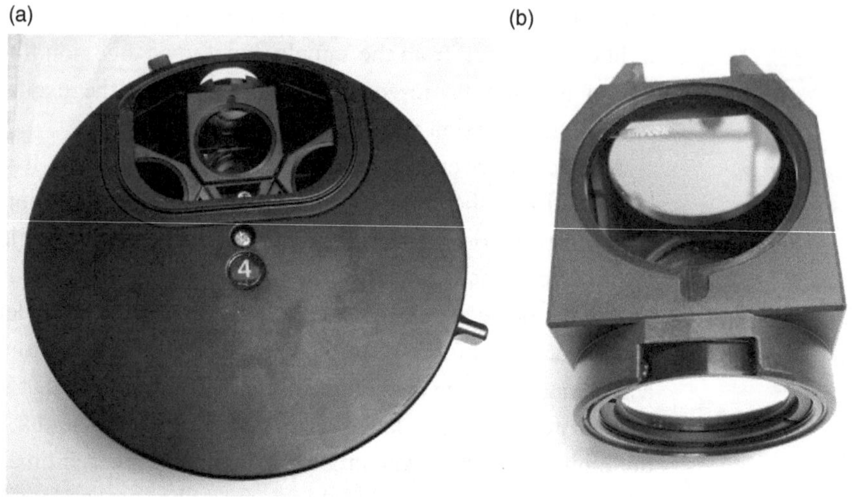

Figure 4.38

(a) Fluorescence filter turret. (b) Fluorescence filter cube.

Strengths and Limitations

The biggest limitation of wide field fluorescence imaging is that samples need to be flat. Any three-dimensionality to the sample will cause blurring in the image due to some of the sample being out of the focal plane of the microscope. This can clearly be seen in the edges of the images in Figure 4.39.

Another issue is that proper sample preparation is critical, and unfortunately is not always straightforward. Remember that at a molecular level, you are essentially trying to hang lights on the structure of interest and you are interpreting the pattern of lights as a real representation of the biological structure. If you did not actually "hang the lights" where you thought you did, you can easily be misled.

What Samples Are Appropriate?

Only very flat samples such as adherent cultured cells or microtomed thin sections of tissue that can be fluorescently labeled are appropriate for wide field fluorescent imaging.

Common Pitfalls

There are a plethora of sample preparation pitfalls, the details of which are beyond this text. In terms of hardware pitfalls, make sure that your light sources, filters, and dyes all match spectrally and that you eliminate as much stray light as possible.

Sample Data

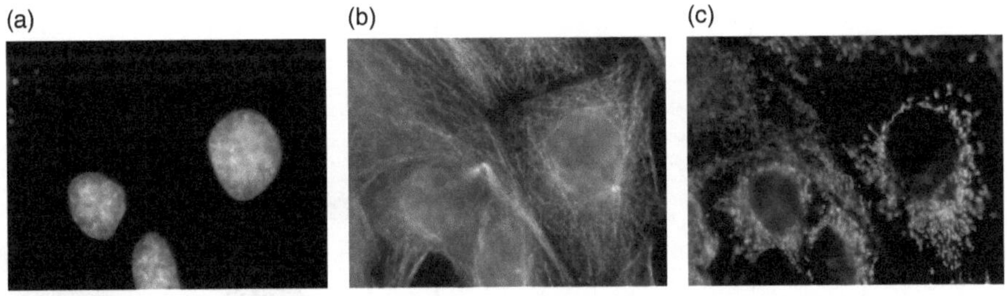

Figure 4.39

Wide field fluorescent images of MDCK cells. (a) DAPI labeling double stranded DNA, highlighting the cell nuclei positions. (b) Alexa 488 immunolabeled microtubules. (c) Mitotracker Red labeled mitochondria.

4.8 Total Internal Reflection Fluorescence Microscopy

Underlying Physical Principles

Total Internal Reflection Fluorescence (TIRF) microscopy uses total internal reflection (Section 1.8) to limit the penetration of the excitation light to ~100 nm depths into the sample. In TIRF microscopy, excitation light is sent up the side (instead of through the center) of a high NA oil immersion microscope objective lens. Then, the light hits the sample at such an angle that at the interface between the coverslip and the sample, the light undergoes total internal reflection. Only an evanescent electric field penetrates into the sample, localizing the excitation of fluorophores to a sample volume that is approximately 100 nm into the sample from the coverslip glass. The fluorescent signal is then collected down the center of the microscope objective, filtered, and sent to a detector, similarly to the wide field fluorescent microscope setup. It is important to note that TIRF only localizes the excitation field in the z-direction. The resolution in x and y stays at the same diffraction limited value of ~$\lambda/2$ or ~250 nm.

Total internal reflection fluorescence microscopy is used to obtain a higher contrast image of the fluorescent molecules that are within 100 nm of the cover glass. The fluorophores in the rest of the sample beyond 100 nm are not excited with TIRF illumination.

What Scientific Questions Can Be Asked?

With TIRF, one can characterize the interaction of fluorescently labeled structures with the cover glass, as well as the interaction between two fluorescently labeled objects, where one is immobilized on the cover glass. For this reason, TIRF is

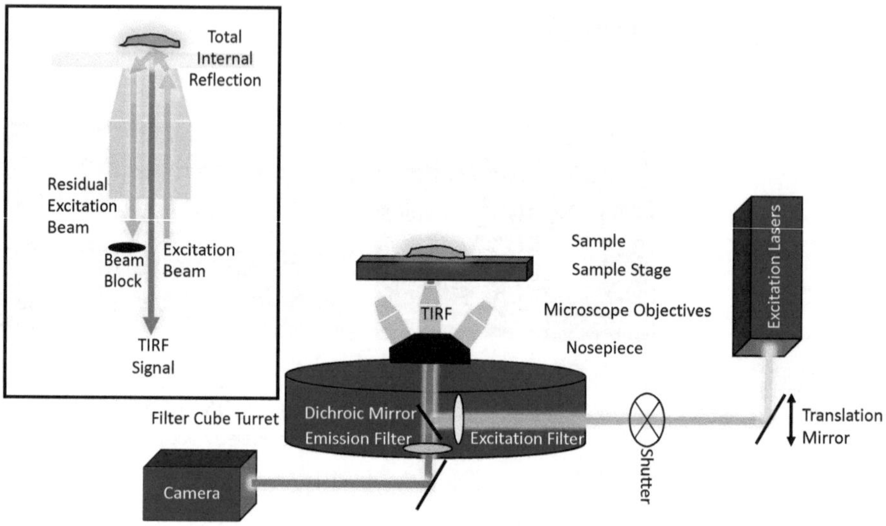

Figure 4.40

Schematic of a TIRF microscope.

often used to characterize focal adhesions, which are the proteins that cells use to attach themselves to objects and surfaces. Total internal reflection fluorescence can also be used for *in-vitro* experiments that characterize the interactions of two molecules, one of which is immobilized on the cover glass while the other is floating in solution, free to bind and unbind with the immobilized molecule. Such an experiment can be used to characterize the kinetics of substrate binding and unbinding for proteins, the interaction of antibodies and antigens, the diffusion rates of molecules in solution, etc.

Strengths and Limitations

The two limitations of the TIRF technique are that the sample must be fluorescently labeled and that the structures of interest must be within 100 nm of the coverslip. Total internal reflection fluorescence is a photon-limited technique since it only excites very few (sometimes single) fluorescent molecules. Therefore, expensive electron multiplying charge coupled device (EMCCD) cameras (Section 2.3.48) are usually required for TIRF microscopy.

What Samples Are Appropriate?

For TIRF, use fluorescently labeled samples that have structures of interest within 100 nm of the coverslip. Alternatively, to characterize the interaction of two fluorescently labeled molecules, immobilize one of the molecules on the coverslip

Figure 4.41

TIRF image of fluorescently labeled microtubules attached to cover glass. Sample courtesy of the Reck-Peterson lab.

and flow a solution of the other labeled molecule past it. Moreover, the sample must be mounted in a low index of refraction mounting media so that it is physically possible to obtain the total internal reflection condition.

Common Pitfalls

Common issues encountered in TIRF arise when the structure of interest is more than 100 nm from the cover glass. Be sure the sample and coverslip are clean. Another issue can arise when the launch angle of the laser is not correct to achieve total internal reflectance. As mentioned above, be sure the index of refraction of the medium that contains the sample of interest is low enough to achieve total internal reflection. Finally, double check that the gain on the EMCCD is set correctly. In low signal-to-noise techniques such as TIRF, the detector must be configured properly to detect low-intensity signals.

4.9 Confocal Microscopy

Real biological samples are usually three-dimensional objects. Throughout the last 70 years, scientists realized that understanding the three-dimensional structure of biological samples is key to understanding their function. Confocal microscopes were developed to do exactly this. The original patent for confocal microscopy was filed in 1957, but the technique did not become widespread until the 1980s when increases in computing power enabled efficient automation and data processing, which are crucial for handling the data sets from confocal microscopy.

Confocal microscopy techniques are most commonly used for fluorescent samples, but can be used for Raman microscopy or even reflected light imaging to build up three-dimensional images of samples. Due to the automation levels needed, confocal systems are usually purchased from a commercial vendor.

Underlying Physical Principles

One limitation of wide field fluorescent imaging is that any fluorescent molecules that are above or below the focal plane of the microscope objective will be blurred, but will still contribute to the background signal of the image, decreasing the contrast and resolution of the images.

In confocal microscopy, a pinhole (also known as a confocal aperture) is introduced as a spatial filter at a position in the optical path to reject light that comes from above or below the focal plane of the objective. A laser beam is focused on the sample and the collected signal light is focused again through the confocal aperture. The signal light can be fluorescent emission, Raman scattering, or even just reflected light. If the signal light comes from the focal plane, it will focus down to the center of the confocal aperture and will be passed on to the detector. If signal light originates from above the focal plane, it will be focused above the confocal aperture and will be diverging again at the pinhole, and thus will be blocked before it gets to the detector. If signal light originates from below the focal plane, it will not be focused by the time it gets to the confocal aperture and will also be rejected by the pinhole. The smaller the pinhole, the thinner the slice in z from the sample that the light will originate from.

In laser scanning confocal microscope systems, the laser beam is scanned in x and y with a set of galvo mirrors to build up a high-contrast image using the confocal aperture to reject background light. The microscope objective can then be moved in z and another x–y image collected. Eventually a three-dimensional image of the sample can be generated from the "z-stack" of images. Laser scanning microscope systems are often called LSM systems.

What Samples Are Appropriate?

Photons must be able to get into and out of the sample. Therefore, there is a limitation to how deeply into a sample confocal microscopy can image that is imposed by how much the incident and emitted light is scattered by the sample, as well as the working distance of the objective.

Figure 4.42

At the heart of all laser scanning techniques is a set of scanning mirrors. One mirror will control the laser's *x* position and another mirror will control the laser's *y* position.

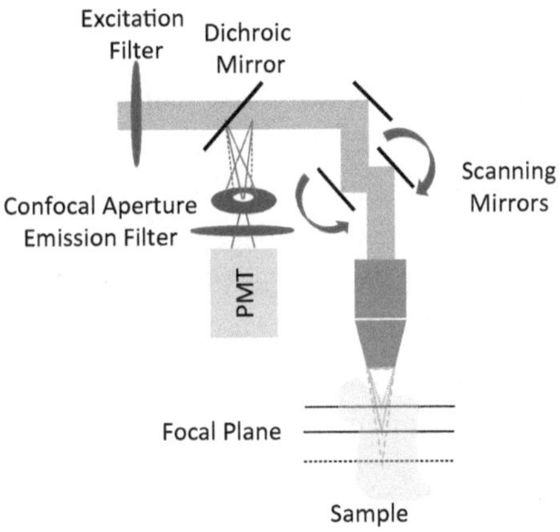

Figure 4.43

Setup for laser scanning confocal microscope.

Limitations

Fluorescent confocal microscopy excites fluorophores that are out of the focal plane, but throws away (with the confocal pinhole) the emitted fluorescent photons that arise from fluorophores out of the focal plane. Since the fluorophores are excited multiple times at each *z* level in the stack, this leads to **photobleaching**, which is the fading of the fluorophores following repeated excitation. Moreover, scanning a laser across a sample and collecting data point by point can be a slow process. Scanning mirrors have been getting significantly faster recently with the advent of resonant scanners,

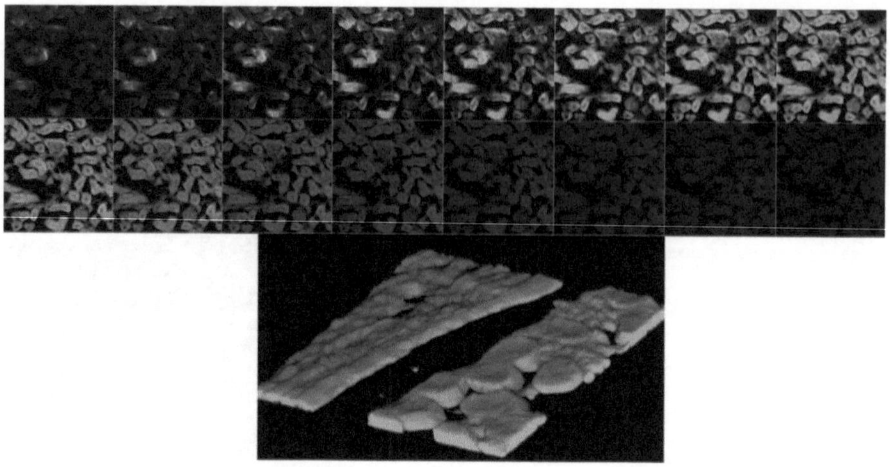

Figure 4.44

Top: confocal *z*-stack of a mouse kidney fluorescently labeled for actin. Bottom: three-dimensional reconstruction of the confocal data. Images courtesy of Dr. Douglas Richardson at the Harvard Center for Biological Imaging.

more sensitive detectors, and brighter fluorescent dyes. Spinning disk confocal microscopes tried to address the speed issue by using two spinning disks with hundreds of pinholes and lenses working in parallel. The improvement in speed of LSM confocals has negated a lot of the spinning disk confocal advantage. LSM systems use PMT or GaASP detectors which are an order of magnitude less expensive compared to the cameras used for detection in spinning disk systems. This means that a typical LSM system can simultaneously collect 3–6 fluorescent channels while a typical spinning disk confocal can simultaneously collect only one or two.

4.10 Light Sheet Microscopy or Selective Plane Illumination Microscopy

Light sheet microscopy is also called selective plane illumination microscopy (SPIM). Light sheet microscopy allows for very fast optical sectioning of samples.

There are commercial light sheet microscope systems available but there is also a widespread homebuilders/do-it-yourself movement around SPIM microscopy. Openspim.org is a site to share designs and tips for those interested in building their own selective plane illumination microscope. Several plugins have been written for ImageJ to handle some of the smaller SPIM data sets. For the open SPIM configuration, samples are mounted in a hydrogel column.

DiSPIM is a variation on SPIM that uses two fixed imaging objectives at right angles to each other. The advantage of the DiSPIM configuration is that samples can stay in a petri dish so that cells and embryos that need to stay in media to grow can still be imaged. Dispim.org is a wiki page for those interested in building their own DiSPIM microscope.

Underlying Physical Principles

In light sheet microscopy, the speed limitations of laser point scanning microscopes are addressed by using a cylindrical lens to focus light incident on the sample to a plane instead of a point. This illuminated plane is then imaged by a camera through an objective lens mounted at 90 degrees to the excitation beam. This configuration allows faster data acquisition since the light sheet illuminates an entire x–y plane at once. The plane illumination can then be swept through the sample in z either by moving the sample or by scanning the position of the laser. This configuration also cuts down on photobleaching of the sample, since only fluorophores that are contributing to the signal are being excited.

Strengths and Limitations

The advantage of light sheet microscopy is fast optical sectioning, meaning it is possible to quickly analyze samples with important three-dimensional features.

The sheer amount of data that can be generated from a light sheet microscope is one of the current challenges for this technique. For example, one cleared mouse brain with volume ~1 cm^3 can be scanned in about three hours on a commercial light sheet system and will generate ~1 terabyte of data! Even transferring a terabyte of data from one hard drive to another is nontrivial, not to mention processing a terabyte of volumetric image data for meaningful results.

As the sheet of light is sent through the sample, the intensity of the fluorescent image on the side opposite the light source will be dimmer. This limitation can be overcome either by illuminating the sample from both sides at once, which can be a bit of an alignment headache, or by rotating the sample and recording data sets with different areas of the sample facing the excitation light and subsequently recombining the data sets in post-processing.

What Samples Are Appropriate?

Samples need to be fluorescent and optically transparent to work well for light sheet microscopy. There has been a proliferation of tissue clearing techniques for making biological samples optically transparent, such as CLARITY, SCALE, BABB, CUBIC, and iDISCO. These clearing techniques are utilized to great

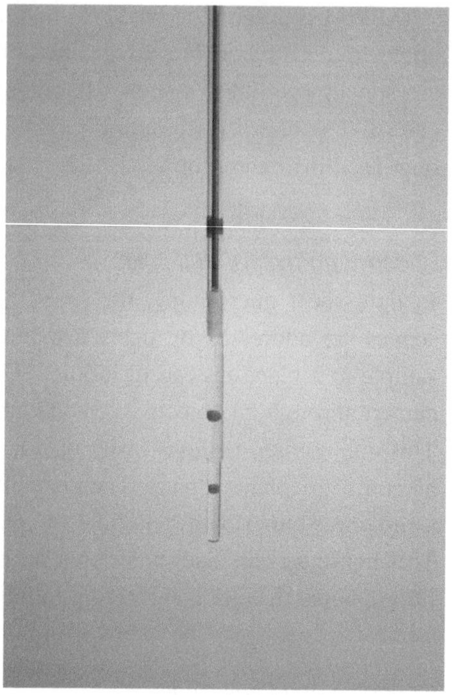

Figure 4.45

Sample mounted for light sheet microscopy in agarose. Image courtesy of Dr. Douglas Richardson at the Harvard Center for Biological Imaging.

Figure 4.46

Cleared mouse brain ready to be mounted for light sheet microscopy. Image courtesy of Dr. Douglas Richardson at the Harvard Center for Biological Imaging.

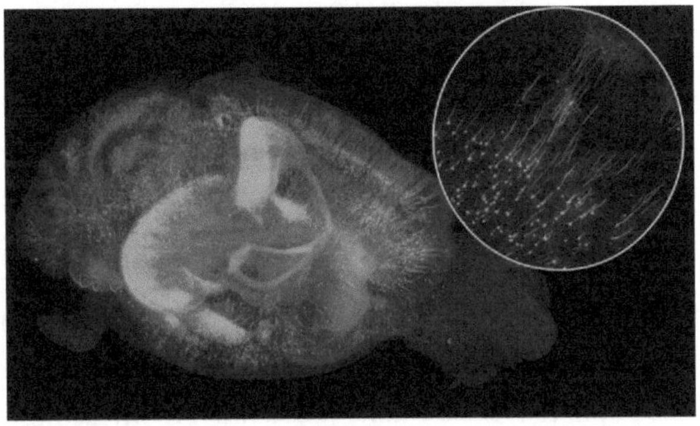

Figure 4.47

Volumetric rendering of light sheet data set of neurons in a cleared mouse brain. Images courtesy of Dr. Douglas Richardson at the Harvard Center for Biological Imaging.

effect with light sheet microscopy. Samples for the openSPIM configuration need to be mounted in an entirely different manner from traditional microscope slides for light sheet imaging since the sample must be illuminated from the side instead of from the top or bottom. Typically, samples are embedded in a cylinder of hydrogel, usually agarose, in a glass capillary tube that is suspended in the light sheet microscope sample chamber. When the light sheet imaging is performed, the agarose is gently pushed out of the tube so that the portion with the sample is hanging under the glass tube, so that the glass tube does not interfere with the imaging. In DiSPIM configurations, samples can be left in media in a petri dish.

4.11 Multiphoton Microscopy

Multiphoton microscopy (commonly abbreviated MP by vendors) is a class of optical microscopy in which more than one photon is used to excite the sample. The most common multiphoton techniques are second harmonic generation (SHG) and two-photon fluorescence (TPF), which are described below. Third harmonic generation is also possible, although less common. Coherent anti-Stokes Raman scattering and stimulated Raman scattering are also multiphoton techniques which are described briefly in Sections 3.5.13 and 3.5.14. Homebuilt multiphoton systems have long been built using a commercial confocal microscope as a base platform. As the limits of multi-photon microscopy are pushed, the commercial systems that offer more robustly engineered solutions are starting to take over in labs.

Underlying Physical Principles

Briefly, all multiphoton microscopy techniques rely on pulsed lasers and nonlinear optics. In a focused picosecond or femtosecond laser pulse, so many photons are pushed into the same time and space that the probability of two (or three) of them interacting with a single molecule becomes a finite reality. The mathematical descriptions of multiphoton phenomena are extensive and beyond the scope of this chapter. But, one key feature of multiphoton microscopy is that the signals generated in these techniques only come from the focal volume, as that is the only location where there are enough photons in the same time and space that multiple photons have a finite probability of interacting with the same molecule at the same time. This fact provides an advantage; a signal is only generated at the focal plane. Therefore, multiphoton processes are inherently synergistic with high-contrast three-dimensional imaging.

Multiphoton refers to the fact that multiple photons must interact simultaneously with the sample to produce the signal. Multiphoton imaging is performed in a laser scanning configuration very similar to the laser scanning confocal microscope (Section 4.9). Specifically, multiphoton microscopy systems generally consist of a pulsed femtosecond laser source for excitation, a set of scanning mirrors (see Figure 4.42) to raster the incident light across the sample, microscope objectives to focus the light, short-pass filters to allow the signal to reach the detector, and photomultiplier tube (PMT) detectors. Since the signal is only generated from the focal spot, the confocal pinhole compared to confocal microscopy can be removed. Like confocal microscopy, z-stacks can be collected to generate three-dimensional data sets.

Often, multiphoton microscopy uses near IR excitation light from a pulsed femtosecond laser to increase the penetration depth into biological samples. (Biological materials are more transparent in the near IR because near IR light scatters less than visible light. Sample clearing techniques mentioned in Section 4.10 can also be employed for multiphoton microscopy.) Note, however, that multiphoton microscopy is still diffraction-limited, so the use of IR light does give up spatial (x–y) resolution in favor of deeper z penetration into the sample as compared to single-photon microscopy techniques like laser scanning confocal.

To address a common misconception, note also that pulsed lasers are *not* simply continuous wave (CW) lasers that have a very fast shutter. The pulsing in ultrafast pulsed lasers arises from constructive and destructive interference of multiple standing wave modes in the laser cavity. This interference effectively redistributes the photons so they are bunched up in time, leading to the extremely high electric fields and photon fluxes in a very small space and a very short amount of time that are required to allow nonlinear optical effects to occur.

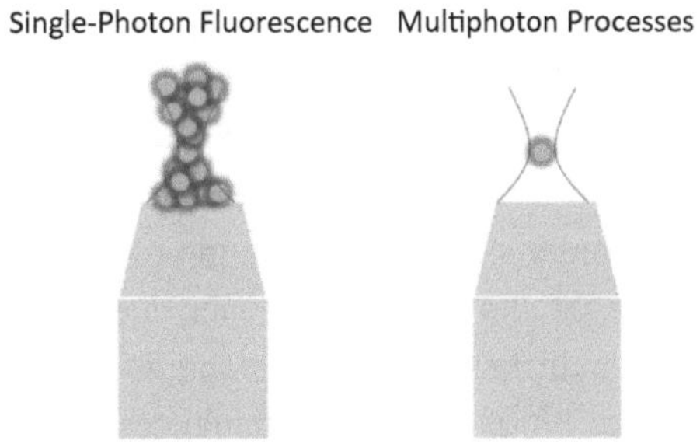

Figure 4.48

Since multiphoton processes require very high photon densities to have a finite probability of occurring, they only generate signal from the focal spot of the laser. In contrast, single-photon fluorescence is generated from the entire laser beam, any time a photon interacts with a fluorophore.

4.11.1 Second Harmonic Generation

Second harmonic generation imaging uses a second-order nonlinear optical phenomenon, in which two photons are simultaneously absorbed and a photon at twice the frequency (twice the energy, half the wavelength) is emitted. It is important to note that the emitted light is *not* dependent on the electronic structure of the sample in question for SHG. There is no energy lost in this process due to relaxation through different energy levels. Often, an ultrafast 800 nm Ti:sapphire laser is used as the light source, and 400 nm light is recorded as the signal on the PMT detector, but a 900 nm excitation photon could be used just as well and the signal would be a 450 nm SHG photon.

Second harmonic generation imaging is one of the label-free imaging techniques. However, it is not universally applicable; SHG only occurs in certain non-centrosymmetric structures, or at surfaces and interfaces where the centrosymmetry is broken by definition.

Collagen fibers are a classic example of a biological structure that will produce SHG (see Figure 4.51). Labeling of the collagen fibers is not required to produce SHG. The intrinsic non-centrosymmetric structure of the collagen allows this multiphoton process to occur. Non-centrosymmetric materials, surfaces, interfaces, or strained centrosymmetric crystals in which symmetry is broken can also generate an SHG signal.

4.11.2 Two-Photon Fluorescence

Two-photon fluorescence occurs when two low-energy photons are absorbed simultaneously by the sample to excite an electron across an energy gap twice the energy of the two input photons. In contrast to SHG, in two-photon fluorescence, the emitted light is dependent on the electronic structure of the sample. In other words, the emission is the same as single-photon excitation fluorescence and may not have exactly twice the energy of the incident beam. The advantage of this technique compared to single-photon excitation fluorescence is that the incident IR light can penetrate deeper into a biological sample. Keep in mind that this is still a diffraction limited technique. Since the excitation light is now in the near IR range (700–1300 nm), spatial resolution will be 350–650 nm, depending on the excitation wavelength used. Thus, the spatial resolution will be worse than single-photon fluorescence imaging by a factor of approximately two.

Despite the deeper penetration of the near IR excitation light, scattering of light in tissue of the emitted light is still a limitation on how deeply into a sample you can image with multiphoton techniques. Microscopists have recently developed two approaches to this problem. First, tissue clearing sample preparation techniques chemically remove some of the proteins and fats that scatter light, rendering the sample optically clear. The other approach uses deformable mirrors (also called adaptive optics) to correct for the scattering by the sample. Ground-based astronomers use adaptive optics to correct for the scattering of starlight due to the Earth's atmosphere. The idea for microscopy is similar; with adaptive optics, one measures the scattering and then bends the deformable mirror into a shape that exactly compensates for the distortion and scattering from the sample.

Two-photon fluorescence can be thought of as very similar to single-photon fluorescence except that two IR photons are used to excite the fluorophore. Since there is a wavelength shift in TPF to lower energies, SHG and TPF signals can be collected simultaneously by using appropriate filters.

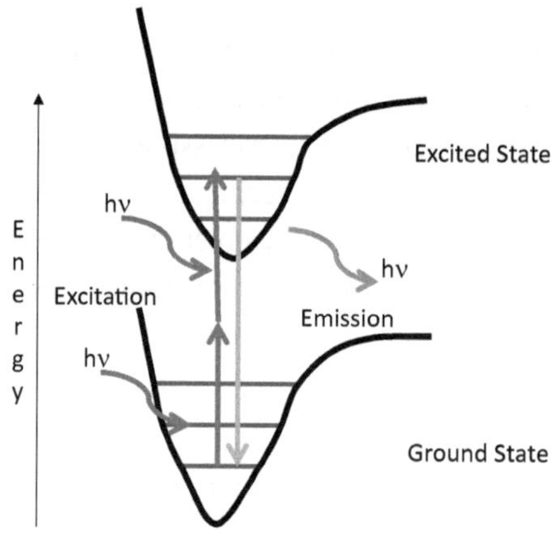

Figure 4.49

Energy diagram for second harmonic generation.

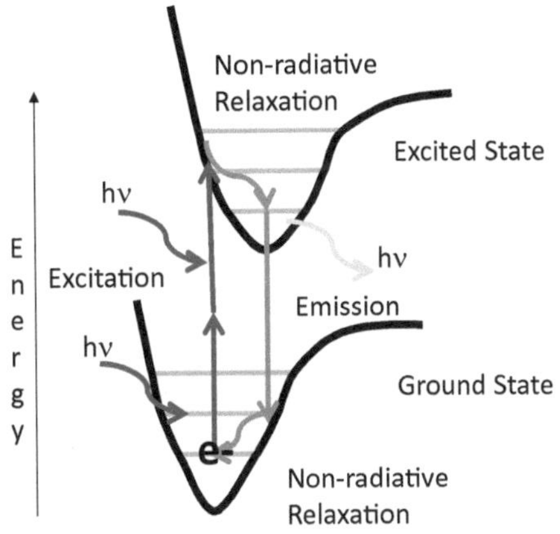

Figure 4.50

Energy diagram for two-photon fluorescence.

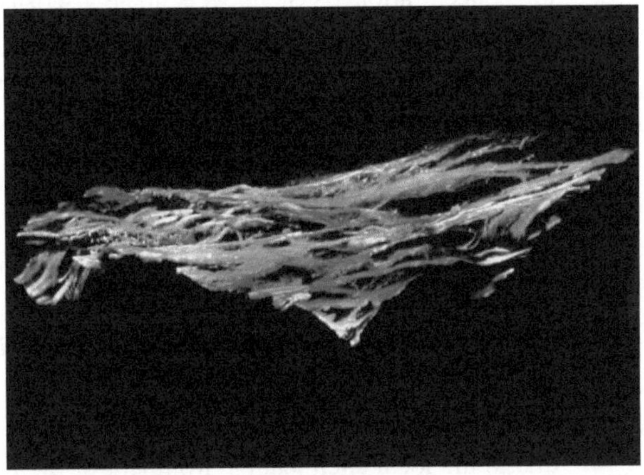

Figure 4.51

Second harmonic generation image of collagen fibers. Image courtesy of Dr. Douglas Richardson at the Harvard Center for Biological Imaging.

4.12 Fluorescence Lifetime Imaging Microscopy/Time Resolved Photoluminescence Imaging/Time Correlated Single-Photon Counting

The time between excitation and light emission from a fluorophore is called the excited state lifetime. The excited state lifetime is different for different fluorophores, and is also sensitive to the local chemical environment of the fluorophore. Measuring fluorescent lifetime can address questions such as what is the local pH,

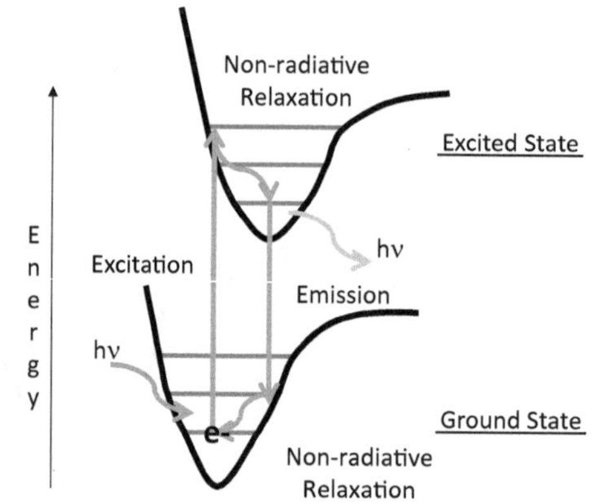

Figure 4.52

Energy diagram of a fluorescent cycle.

ion concentration, oxygen levels, local viscosity, protein binding and aggregation, local refractive index, or local solvent polarity within a subcellular structure.

In the world of solid state physics and materials science, fluorescence lifetime imaging microscopy (FLIM) is known as time resolved photoluminescence (TRPL) or sometimes as time correlated single photon counting (TCSPC).

Underlying Physical Principles

An electron will stay in the excited state for a certain amount of time before it emits a photon and relaxes back down to a ground state. The amount of time between when the molecule absorbs the excitation photon and when the molecule emits a photon is called the fluorescence lifetime. The fluorescence lifetime is typically on the order of a couple nanoseconds, but the exact lifetime can change due to the local chemical environment of the molecule.

4.12.1 Time Domain FLIM

The easiest way to think about setting up such a system is sending in a short laser pulse (picoseconds or femtoseconds in length) toward a sample, and using a set of scanning mirrors (as in confocal microscopy; Section 4.9) to probe each position within the sample sequentially. As each position in the sample is excited, the emission light can be sent to a fast photodetector such as an avalanche photodiode (Section 2.3.41) connected to TCSPC electronics that can monitor the number of photons as a function of time. At each point in the sample, an exponential decay of the fluorescence as a function of time can be determined.

Time domain FLIM can be performed on a conventional confocal microscope (Section 4.9) by changing to a pulsed excitation source and fast photodetectors.

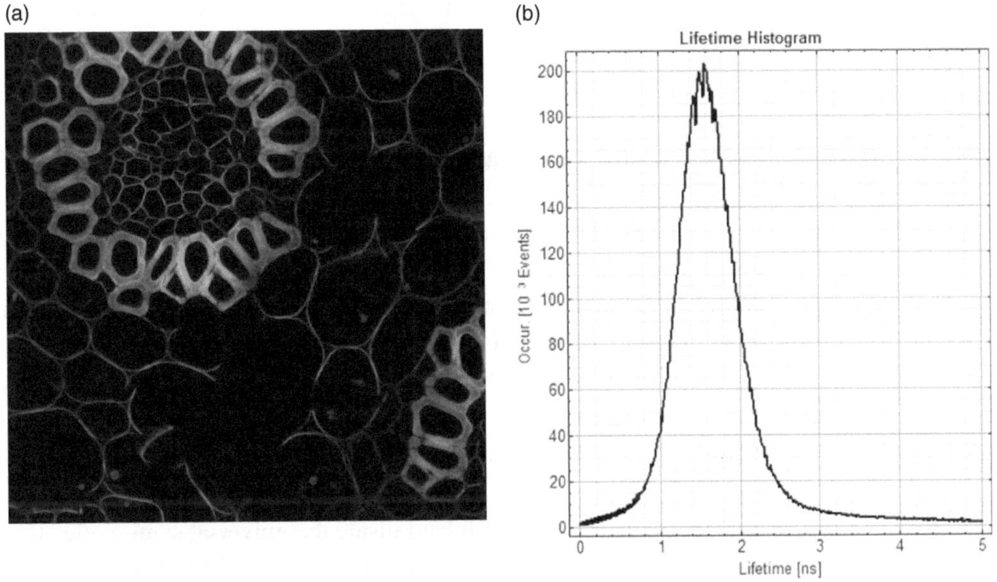

Figure 4.53

(a) Example of time domain FLIM data set of *Convallaria* 500–550 nm emission. (b) Lifetime histogram of *Convallaria* 500–550 nm emission. Images courtesy of Dr. Douglas Richardson at the Harvard Center for Biological Imaging.

Alternatively, time domain FLIM can also be performed on a TPF microscope (Section 4.11) by simply changing the detector to a fast photodetector, since the excitation light is already pulsed. FLIM on a confocal microscope or a two-photon microscope works well with three-dimensional samples by constructing a *z*-stack.

4.12.2 Frequency Domain FLIM

Fluorescence lifetime imaging microscopy can also be performed in a wide field configuration, which can speed up data acquisition significantly by allowing you to collect data in parallel from multiple sample positions. For FLIM in a wide field configuration, use a modulated LED light source and an intensified CCD camera. A series of wide field images (~8–12 images) are collected at different phases of the modulation of the excitation light. The delay in the phase of the emission light is calculated from the wide field images and a fluorescence lifetime can be calculated for each pixel in the image.

Frequency domain FLIM can also be performed on a spinning disk microscope with a modulated laser source and intensified CCD to eliminate some of the out-of-focus light.

Frequency domain FLIM records multiple measurements on each pixel, so the data can be more robust.

4.12.3 Phasor Analysis of FLIM Data

If you have ever used multi-exponential component fitting, you have probably found yourself asking: "How many components should I include in the fit to my data?" With enough exponential components, you can fit almost anything, but then you need a physical explanation for each component.

Phasor analysis takes the Fourier transform of the FLIM data and then plots the real and imaginary parts against each other. If the fluorophores yielding fluorescence in the data have a single lifetime, the point will land on the universal semicircle. If the fluorophores yielding fluorescence in the data have multiple lifetimes, then the point will land inside the universal semi-circle. The Laboratory for Fluorescence Dynamics at the University of California, Irvine, headed by Enrico Gratton has a software package called Global for download to do phasor analysis. No *a-priori* assumptions about the number of exponential components are needed. Global uses a relatively low computational load, so the analysis can be done quickly for every single pixel in the image. Lambert Instruments also has included phasor analysis in its FLIM analysis software.

4.13 Introduction to Super Resolution Microscopy

Light can only be focused to a diffraction-limited spot, which is roughly half the wavelength of the light you are using. For the sake of discussion, we'll approximate the diffraction limit to ~250 nm in the visible light regime. Super resolution techniques allow you to resolve structures smaller than 250 nm. To date, all optical super resolution techniques require a fluorescently labeled sample.

We can image at better than the diffraction limit in two ways. First, we can perform post-processing on data. The diffraction limit is really a way of saying that a single point of light will be blurred by the point spread function (PSF) to a larger spot. Since the PSF can be measured, it is possible to remove its blurring effect in post-processing with deconvolution algorithms (Section 4.14).

Second, there are several approaches to "beating" the diffraction limit on the data-collection side of microscopy too. This is an active area of research right now and new techniques are announced periodically. None of the techniques break the laws of physics. It is important to understand the strengths and limitations of each of the approaches. Point spread function engineering and localization of fluorophores are the two optical approaches to visualizing structures that are smaller than 250 nm. Again all of these approaches require the sample to be fluorescently labeled.

Point spread function engineering on the data-collection side shapes the excitation light to extract information from an area that is smaller than the diffraction limit of light. The two most common techniques in this category are stimulated emission depletion (STED) microscopy (Section 4.15) and structured illumination microscopy (SIM; Section 4.16).

Alternatively, we can beat the diffraction limit experimentally by localizing fluorophores. Localization of fluorophores involves only a sparse subset of the fluorophores in a sample emitting at any given time. An image of the sparse set of fluorophores is captured and then the diffraction-limited spots of light are fit to a Gaussian function. The center of the Gaussian fits can be determined to ~20–30 nm localization accuracy. The center positions of the fits are then recorded on a map of the sample. This process is repeated a number of times with different fluorophores to build up a super resolution image. The localization techniques are called stochastic optical reconstruction microscopy (STORM) or photo activated localization microscopy (PALM; both discussed in Section 4.17). There are a number of variations on each of these techniques.

If your samples are not fluorescently labeled, consider scanning probe techniques or electron microscopy. Scanning probe techniques are a family of techniques that use an atomically sharp tip to interact with the sample. They offer superb spatial resolution and can provide information down to the atomic level. Atomic force microscopy is the most common. A brief overview is given in Section 4.18. There are several approaches to combine optical spectroscopy techniques with scanning probe techniques (Sections 3.4.15 and 3.5.16).

Electron microscopy uses electrons instead of photons for imaging. Electron microscopy includes two primary methods. First, scanning electron microscopy (SEM) scans an electron beam back and forth across a sample and collects the back-scattered or secondary emission electrons, somewhat analogous to a laser scanning microscope in reflection mode. Second, in transmission electron microscopy (TEM), the electrons pass through the sample, analogously to transmitted optical microscopy techniques such as bright field and dark field microscopy. A very brief overview of electron microscopy techniques is given in Section 4.19. SEM can achieve spatial resolution down to ~1 nm, and SEM images can take a couple seconds to a couple minutes to record, depending on your scan speeds. Transmission electron microscopy can have spatial resolution as low as 0.08 nanometers. Resolving single atoms in an aberration corrected TEM or aberration corrected scanning TEM system is now possible on a somewhat routine basis.

4.14 Deconvolution

Deconvolution refers to the mathematical processing of a fluorescent image to remove the blurring caused by the PSF. Deconvolution functions can be applied to data with a number of software suites.

Underlying Physical Principles

Consider a point source of light on a microscope sample stage (e.g., a 20 nm fluorescent bead). The image of the 20 nm bead will be blurred out to ~250 nm by the PSF. If you take a series of images with different *z* levels (a *z*-stack) of the single 20 nm fluorescent bead, you can measure the PSF for the microscope system, which is the mathematical description of exactly how much the light from each fluorophore is being blurred by the real-world limitations of the optics. Deconvolution then mathematically removes the blurring by the PSF to get a clearer "super resolution" image. A blind deconvolution technique doesn't require the measurement of the point spread function, as it relies on some assumptions about the blurring of the light and then adapts the assumptions about the PSF as it iteratively tries to improve the sharpness of the image.

Deconvolution can be applied to most microscopy techniques, including wide field fluorescence, confocal, light sheet, and multiphoton microscopies. There are a number of deconvolution algorithms, the details of which are far beyond the scope of this book but can be found in other textbooks listed at the end of this chapter and in open-source codes widely available on the internet. The biggest challenge for deconvolution has been the computer processing power needed, but with the drop in the cost of cloud computing and processors, access to computer processing of optical imaging data is becoming more widespread.

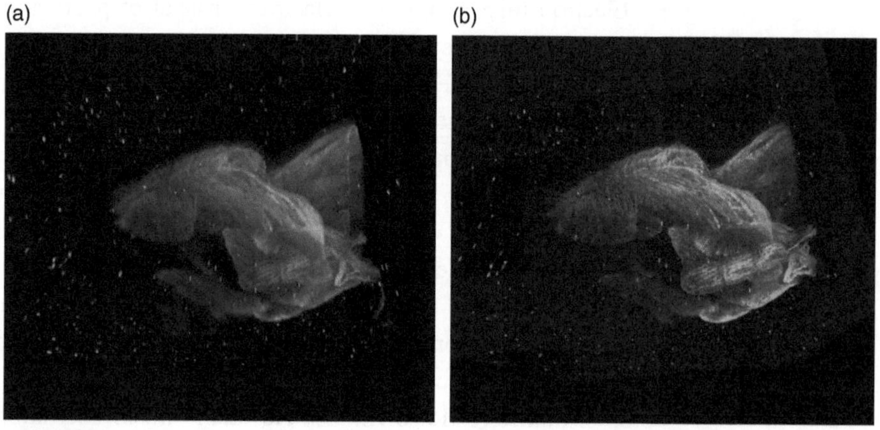

Figure 4.54

(a) Light sheet microscopy data set with no deconvolution algorithms applied. (b) Light sheet microscopy data set with deconvolution applied. Images courtesy of Dr. Douglas Richardson and Dr. Liz Sefton.

4.15 Stimulated Emission Depletion Microscopy

Stimulated emission depletion (STED) microscopy was the first of the "super resolution" techniques to be demonstrated in 1999. It kickstarted something of a renaissance in optical microscopy technique innovation that continues today to push imaging beyond the diffraction limit. Sample prep for STED microscopy is similar to standard fluorescent sample prep but with a higher labeling density.

Underlying Physical Principles

Fluorescent molecules can be stimulated to emit a photon and transition back to the ground state with light. In STED, two overlapping femtosecond lasers are scanned across the sample. The first laser beam excites a diffraction-limited spot. The second beam is shaped like a torus and is just slightly time delayed compared to the first beam. The toroidal beam stimulates the emission of the fluorophores around the edge of the area that was excited by the first beam, leaving a sub-diffraction-limited area still excited. The spontaneous fluorescence from the sub-diffraction-limited area is then collected a couple nanoseconds later as the super resolution signal.

One of the biggest challenges for STED is photobleaching, as you need to send each fluorophore through the fluorescent cycle multiple times as you scan your excitation and stimulated emission beams across the sample. Stimulated emission depletion systems can be homebuilt on a confocal setup or are commercially available. The commercial systems continue to get easier to implement and use. For instance, Abberior Instruments has recently released an impressive STED module that can be mounted on most microscope bodies, named STEDYCON.

4.16 Structured Illumination Microscopy

Structured illumination microscopy (SIM) is probably the most straightforward of the super resolution techniques. (Care should be taken when using the abbreviation

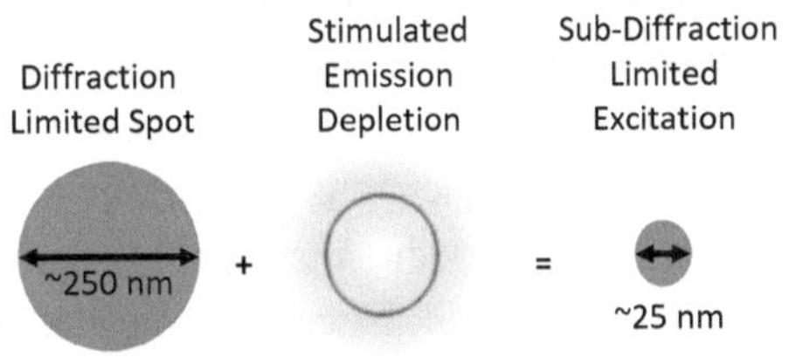

Figure 4.55

In STED microscopy, a torus-shaped stimulated emission beam de-excites the fluorophores in the outside edge of a diffraction-limited beam, leaving just the fluorophores in a sub-diffraction-limited spot excited.

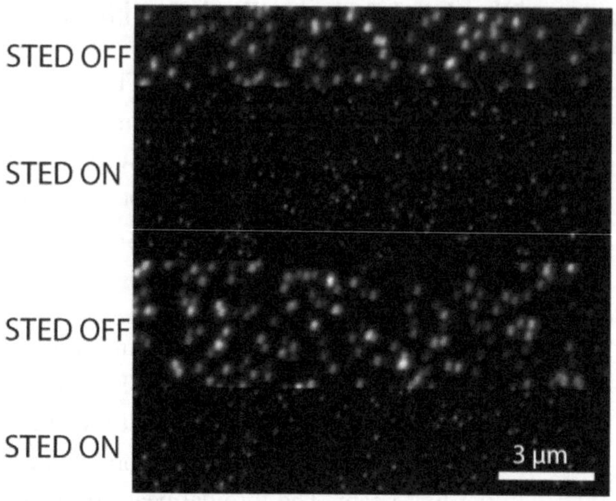

Figure 4.56

Demonstration of the improved spatial resolution of stimulated emission depletion microscopy. The sample was yellow/green 40 nm fluorescent beads imaged with 490 nm excitation and 590 nm STED lasers. The STED laser was turned on and off to show the enhanced spatial resolution. Image is courtesy of Dr. Douglas Richardson.

Figure 4.57

A one-dimensional periodic image.

SIM when writing or speaking to other scientists; it can easily get confused with SIMS, which is secondary ion mass spectroscopy.) Sample prep for SIM is similar to standard fluorescent microscopy sample preparation.

Underlying Physical Principles

Structured illumination relies on Fourier optics. In Fourier optics, an image can be thought of as a series of spatial frequencies adding up to the image. Let's start by considering a very simple, one-dimensional periodic image (Figure 4.57).

A spatial frequency is a cycle per unit length. Let's say each block in the one-dimensional periodic image has length of 0.5 units. So, one cycle (one light block and one dark block) is completed in 1 unit length. Fourier proposed that any pattern could be described by an infinite series of sine or cosine waves. We could start trying to describe our one-dimensional image with just a single sine wave:

$$\frac{4}{\pi}\sin[2\pi x] \tag{4.1}$$

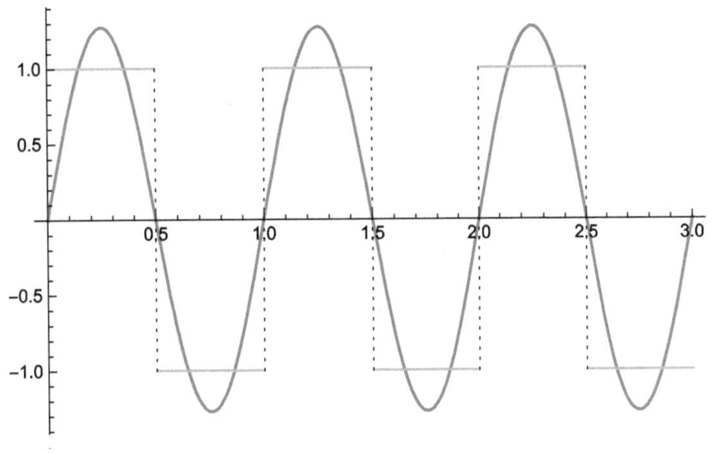

Figure 4.58

A single sine wave fit to the one-dimensional image in Figure 4.57.

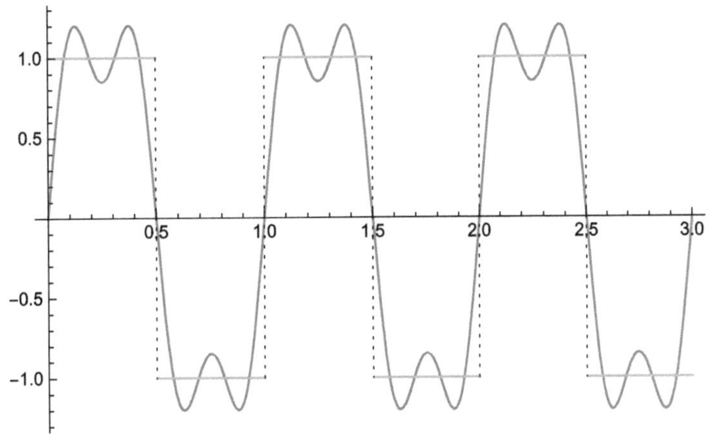

Figure 4.59

A two sine wave fit to the one-dimensional image in Figure 4.57.

The fit improves if we add in another sine wave with a higher frequency.

$$\frac{4}{\pi}\left(\sin[2\pi x] + \frac{1}{3}\sin[6\pi x]\right) \tag{4.2}$$

The fit improves as we add more higher frequency components.

$$\frac{4}{\pi}\left(\sin[2\pi x] + \frac{1}{3}\sin[6\pi x] + \frac{1}{5}\sin[10\pi x] + \frac{1}{7}\sin[14\pi x] + \frac{1}{9}\sin[18\pi x] + \frac{1}{11}\sin[22\pi x] + \frac{1}{13}\sin[26\pi x]\right) \tag{4.3}$$

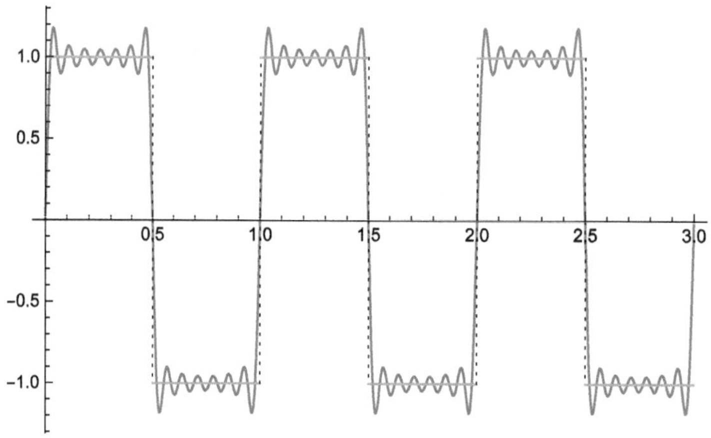

Figure 4.60

A seven sine wave fit to the one-dimensional image in Figure 4.57.

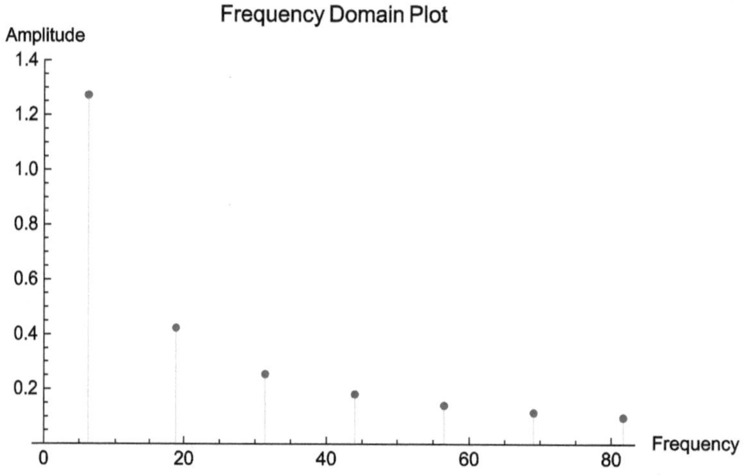

Figure 4.61

Frequency domain plot of component sine waves fit to the one-dimensional image in Figure 4.57.

If we took our one-dimensional image and plotted the amplitude and frequency of the different sine wave components that we were using to build up the image, we would have transformed the image from the spatial domain (also known as "real space") to the frequency domain (also known as "k-space" or "Fourier space").

Despite the fact that the two plots look totally different, they contain exactly the same information. In fact, you can convert between the frequency domain and the spatial domain at will. This concept can be extended to a two-dimensional image:

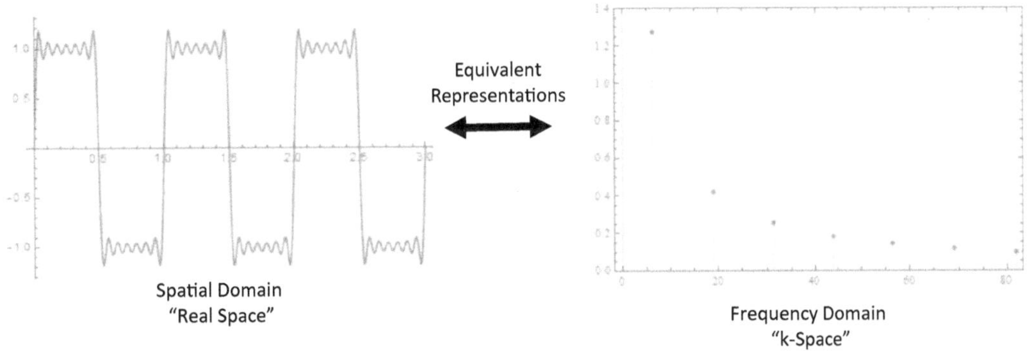

Figure 4.62

Using a Fourier transform, it is possible to transition from the spatial domain to the frequency domain and back.

a series of sine wave patterns in two dimensions added together describes the full image.

Let's switch the discussion to real space (spatial domain) for a quick reminder. Ernst Abbe described in his theory of image formation that it is necessary to collect the first-order diffracted light in order to generate an image.

Part of the resolution limit of any microscope depends on the numerical aperture, which describes the angle of light that a lens can collect. Smaller structures diffract light at higher angles. So to generate an image of small structures, a wider angle of collection is necessary.

Now back to Fourier space: the axes of the frequency domain (Fourier space) image are k_x and k_y, where k is the wave vector along a certain direction. This is why Fourier space is sometimes referred to as k-space. The low-frequency components are at the center of the image. The high-frequency components are at the edge of the image.

The focal plane and the image plane of a lens are different. The image plane is necessarily in real space since it corresponds to your sample. For reasons beyond the scope of this book, the focal plane is actually the Fourier transform of the image plane.

However, the back focal plane of the microscope objective is finite in size, so it effectively acts as a spatial filter, putting a hard limit on the highest frequencies that can be transmitted through the microscope. The high-frequency components (i.e., those at the edges of the back focal plane) help represent the sharp edges that we observed modeling the square wave in one dimension.

Structured illumination microscopy bypasses the limit imposed on the resolution by the finite size of objective lenses' back focal plane. In SIM, the sample is illuminated with a known pattern of light to generate Moiré fringes (Figure 4.65)

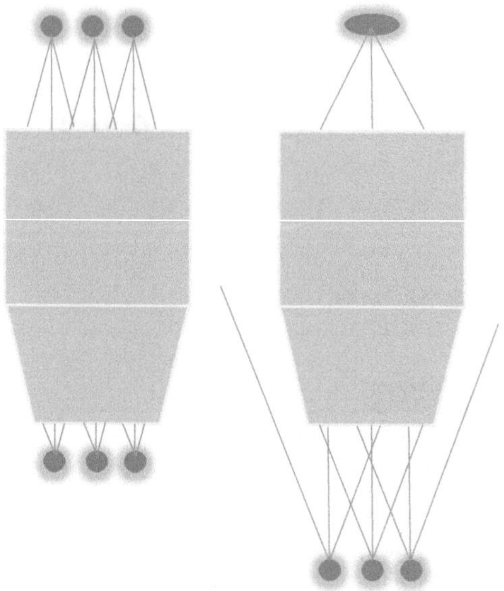

Figure 4.63

To generate an image, the first-order diffraction must be collected.

of lower frequency that are closer to the center of the Fourier plane that will pass through the microscope objective. Then an algorithm calculates what higher-frequency components your sample really contained and reassigns them to the appropriate edges in Fourier space. A Fourier transform is taken to bring the image back to real space, with the new higher-frequency components leading to a higher spatial resolution in the image.

In other words, SIM takes advantage of Moiré fringes to shift the diffracted light from small features back to a lower angle that the microscope objective can collect.

In SIM, the excitation light is patterned into a series of lines. The series of excitation lines generates Moiré fringes as the light interacts with the sample, including the structures that are below the diffraction limit of light. The Moiré fringes are at lower frequency so they can be collected by the microscope objective. The interference lines are directional, so they can only be used to enhance the resolution along that direction. Multiple images must be taken, where for each image the pattern of excitation light is rotated and shifted. To generate the final SIM image, about 27 individual images – each with different portions of the sample illuminated by the grid and different orientations of the grid to generate Moiré fringes along different axes of the image – need to be acquired and processed. This puts a moderate constraint on how fast SIM images can be generated.

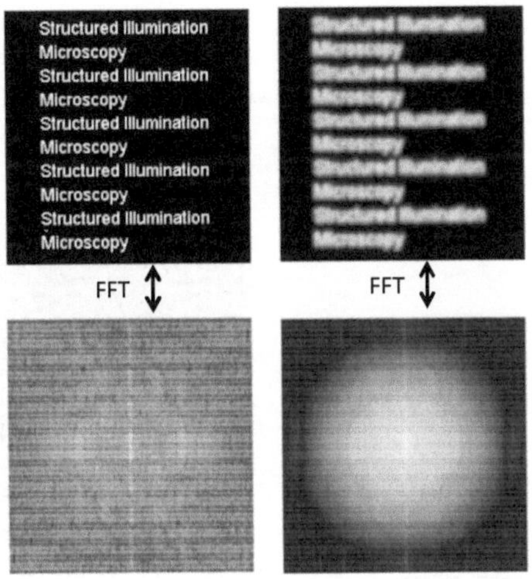

Figure 4.64

High-frequency spatial components are needed to make a sharp edge in an image. In the Fourier plane, the further the point is away from the center, the higher the frequency components are. In a microscope, the Fourier plane is the back focal aperture, so the microscope objective ends up acting as a low-pass filter, cutting out the higher-frequency components and blurring the image. On the left is a sharp image and the fast Fourier transform (FFT) to show the same information in the spatial frequency domain. On the right is a blurred image and the corresponding frequency domain image. Note that the high-frequency components near the edge of the frequency domain image are missing.

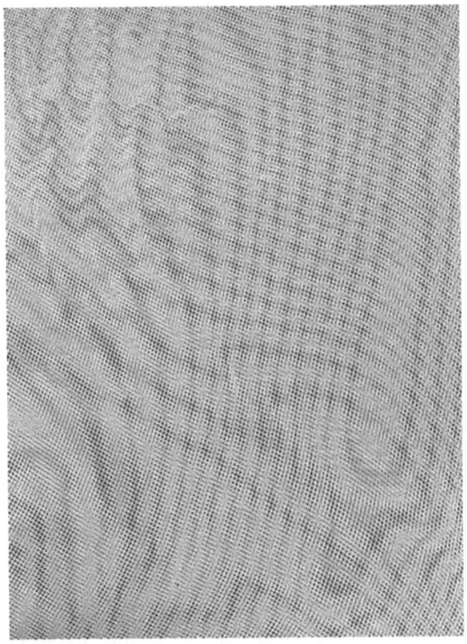

Figure 4.65

Image of Moiré fringes at a macroscopic scale. By overlaying two coarse fabrics, Moiré fringes can easily be observed. The periodic patterns of the fabrics interfere and generate the dark Moiré fringes at a lower spatial frequency than the spacing of the fabric.

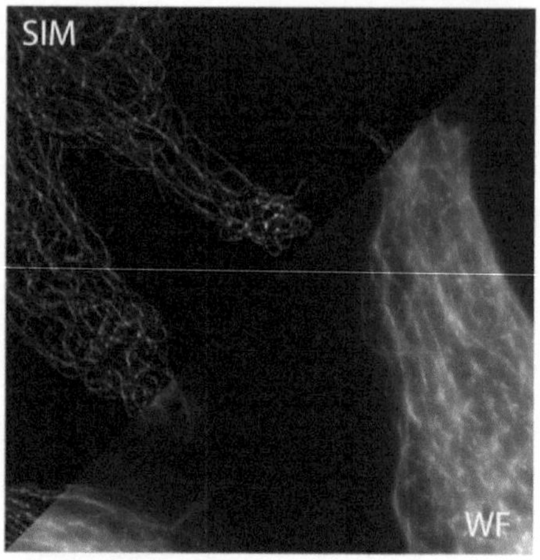

Figure 4.66

Comparison of structured illumination microscopy (SIM) – upper left – and simulated wide field (WF) microscopy – lower right – of fluorescently labeled tubulin in bovine endothelial cells. Image courtesy of Dr. Douglas Richardson at the Harvard Center for Biological Imaging.

Structured illumination microscopy can resolve structures of ~100 nm. This doesn't sound so impressive at first, but comparing the images in Figure 4.66 shows the effectiveness of the technique. The limiting factor remains which spatial frequencies can be collected by the microscope objective. Structured illumination microscopy works with standard preparation of fluorescently labeled samples on a microscope slide.

4.17 Stochastic Optical Reconstruction Microscopy and Photoactivated Localization Microscopy

Underlying Physical Principles

If two fluorophores are closer than the diffraction limit of light, we will not be able to resolve the two fluorophores. In other words, we will not be able to tell if the fluorescent light we detect comes from one or two fluorophores. Remember, however, that the detection limit and the resolution limit are distinct concepts. We can detect a single fluorophore with sensitive detectors such as EMCCDs. We also know that light from a point source will be blurred into a Gaussian shape by the PSF. If we detect a single photon, we can fit the shape of the blurred light and localize the fluorophore down to about 20 nm resolution. The precision of the fit localization depends on the number of photons detected from the single fluorophore. One solution would be spacing out the fluorophores enough so that two fluorophores are never closer than the diffraction limit of light, allowing us to fit the light from the single fluorophores and know their precise location to within ~20 nm of uncertainty. However, this results in low signal to noise. We need dense spatial labeling of fluorophores in a structure to clearly see its shape. But then inherently, this high fluorophore density imposes overlap in the light detected from fluorophores that are spatially close.

Stochastic optical reconstruction microscopy (STORM) and photoactivated localization microscopy (PALM) are two super resolution techniques that use special sample preparation to turn off most of the fluorophores at a given time. If only a few fluorophores are emitting light at once, an image can be captured and the light from the single fluorophores can be fit and the fluorophore localized. Those fluorophores are then turned off and a new set of fluorophores is turned on and the process is repeated for ~10,000 images. The location of each localized fluorophore is compiled to form the final "super resolution" image.

STORM and PALM are usually performed using a TIRF or near-TIRF microscope configuration. The data can be collected in "3D," with a z-range of several hundred nanometers when an astigmatic lens is introduced to the excitation beam. If a fluorophore is right at the focal plane of the image, it will render as a circle. If the fluorophore is above or below the focal plane, the image will be elongated vertically or horizontally, respectively. The amount of elongation is proportional to the distance above or below the focal plane and can be calibrated with a fluorescent bead.

Samples for STORM or PALM must be specially prepared to force the fluorophores to blink. Typically the samples are prepared in a glass-bottomed chamber and an "imaging buffer" is added with some components to help the fluorophores blink. Systems can be homebuilt or commercially purchased. The commercial vendors have published sample preparation protocols which are a useful starting point for understanding the sample preparation requirements.

Common Pitfalls

As with all the optical super resolution techniques, your fluorescent labeling must be much denser than for standard wide field fluorescent or confocal techniques. It is worth screening your samples with a normal fluorescent microscope before you attempt a super resolution technique. If your samples do not generate a good fluorescent image with a normal fluorescent microscopy technique they will certainly not produce a valuable image in a super resolution technique.

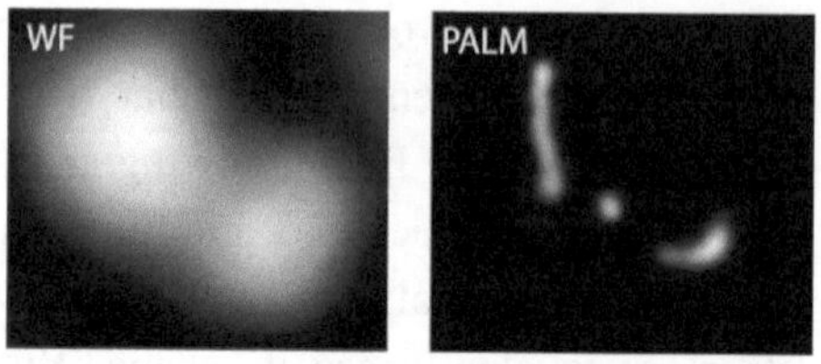

Figure 4.67

Comparison of wide field (WF) microscopy and photo activated localization microscopy (PALM) of mEOS fluorescent protein fused to a member of the focal adhesion complex. Images courtesy of Dr. Douglas Richardson at the Harvard Center for Biological Imaging.

4.18 Atomic Force Microscopy: A Brief Overview

Atomic force microscopy (AFM) is not an optical technique. It belongs to the scanning probe microscopy family of techniques. This section is included in this book as a number of emerging techniques combine optical spectroscopy with atomic force microscopy as another way around the diffraction limit of light. These efforts are noted in the relevant spectroscopy sections (3.4.15 and 3.5.16).

Underlying Physical Principles

In AFM, an atomically sharp probe (~10 nm is a typical tip radius) is mounted on the end of a cantilever. Atomic force microscopy can operate in quite a number of modes. The two most common modes are contact mode and tapping mode. In contact mode, the tip is dragged along the surface of the sample and bounces up and down depending on the surface topography of the sample. A laser beam is reflected off the back of the cantilever. The position of the reflected laser beam is tracked with a quadrant photodiode, which comprises four photodetectors placed in a square pattern. The laser beam is centered at the corner where the four photodiodes meet. By tracking the relative difference between the intensity of the light on the different photodiodes, the position of the laser and hence the height of the AFM cantilever can be monitored.

Dragging a sharp needle across a soft sample is clearly not desirable as the tip will just scratch up the sample. Moreover, the sharpness of the tip determines the spatial resolution of AFM. Dragging the tip across the sample will quickly dull the tip. Tapping mode addresses these challenges. In tapping mode, the AFM tip taps its way across the surface, bouncing up and down at a known resonant frequency depending on the stiffness and other properties of the tip and the AFM instrument. Tapping mode produces the same topography information about the sample, but with less sample damage, and preserves the tip for a longer time. Tapping mode can also provide a phase image. In a phase image, the deviation in the tip frequency from the resonant frequency is tracked, and this can distinguish between materials of different stiffness.

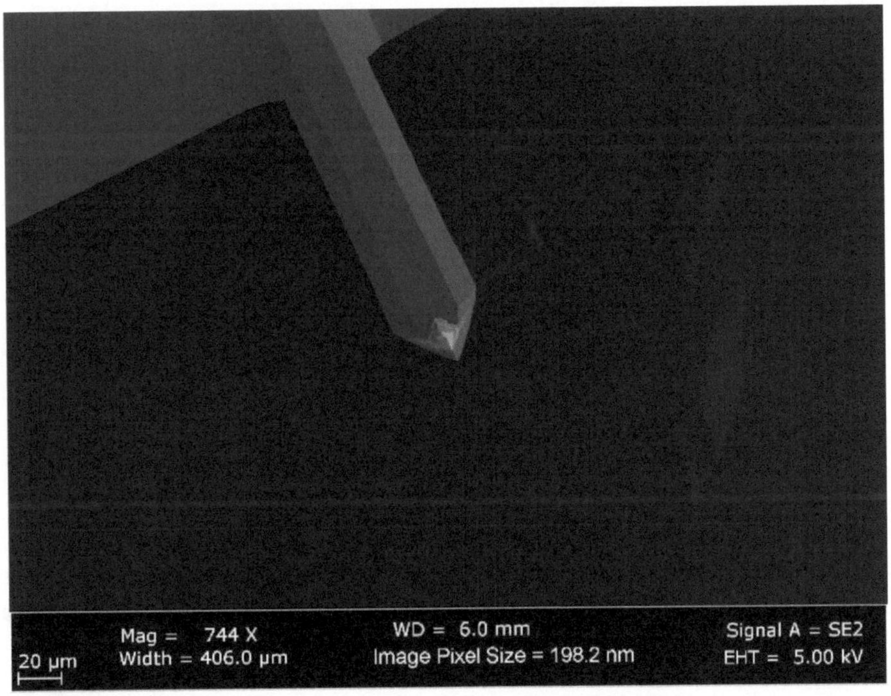

Figure 4.68

Scanning electron microscopy image of an AFM tip. The cantilever sticks out from a chip in the top left of the image. The low-contrast pyramid at the end of the cantilever is the actual AFM tip (pointing up toward the viewer in this image) that interacts with the sample.

What Scientific Questions Can Be Asked

Basic AFM reveals the topography of your sample. Nanomechanics can also be probed with an AFM system using a specialized head called a nanoindenter.

Strengths and Limitations

The strength of AFM is its spatial resolution. AFM allows incredible spatial resolution. Cutting-edge AFMs today can image the atoms in a single molecule. Observing single atomic layer step heights is somewhat routine in the AFM world.

The maximum scan (field of view) area for an AFM is typically less than 100 μm, so the structures on the sample need to be small. In a typical AFM, the maximum sample size that can be mounted is often less than 25 mm in diameter. Atomic force microscopy is a scanning technique. It collects data one point at a time so it can sometimes take hours to collect a single image. The images are an "aerial" view. You can only see the sample from the top. If you have some sort of vertical structure, you cannot visualize anything on the side of the vertical structure.

Atomic force microscopes are very sensitive to vibrations, so they should be placed on an active vibration isolation table inside an acoustic hood. Otherwise your data will include artifacts from the lab door slamming shut and your lab mate laughing loudly at a joke. Information from an AFM is limited to only the sample surface. In standard AFM, there is limited chemical information. (Phase imaging from tapping mode can provide some contrast between materials with different stiffness in a sample.) Without chemical information though it is sometimes hard to say for sure whether a feature is dust or a protein of interest on the sample surface.

Atomic force microscope tips start at around $30 a piece and rapidly increase in price depending on how fancy or specialized the tip is. The tips are consumables and are easy to break with mishandling.

What Samples Are Appropriate?

Samples need to be relatively smooth and small to work well for AFM. While the AFM is sensitive to sub-nanometer height differences, the *z* range of the scan head is often limited to a few microns of roughness.

Common Pitfalls

Aligning samples in AFM can be challenging. The measurements are also quite sensitive to anything that mechanically affects the sensitive tip, including vibrations from equipment or nearby subway trains, changes in temperature in the room, molecules and contaminants absorbed on the sample surface or the tip, users breathing on the equipment, etc.

Schematics

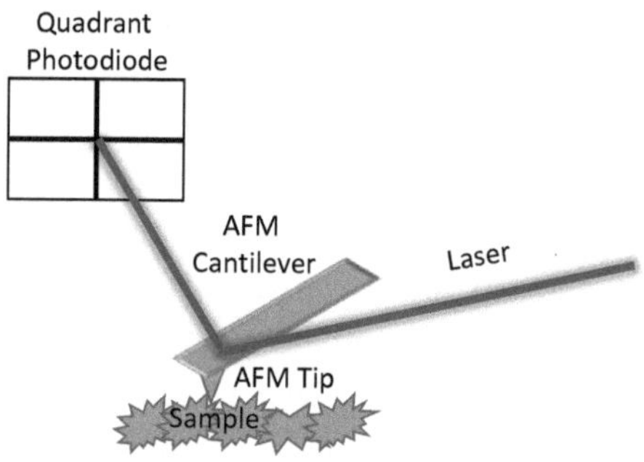

Figure 4.69

Schematic of a typical atomic force microscope system.

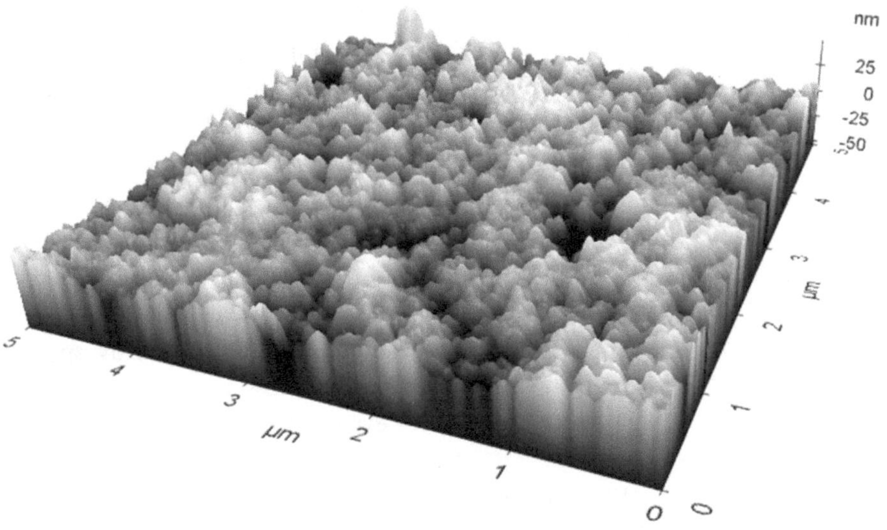

Figure 4.70

Atomic force microscopy image of high-gloss inkjet paper. Note that the spatial resolution of AFM can be sub-nanometer in *x*, *y*, and *z* dimensions. Surface roughness on the nanometer scale is a common AFM measurement.

Sample Data and Data Interpretation

At the basic level, AFM provides images with sub-nanometer resolution in *x*, *y*, and *z* dimensions over the area of the sample that the tip scans. Advanced AFM techniques are beyond the scope of this book.

4.19 Electron Microscopy: A Brief Overview

Electron microscopy (EM) is a huge and complicated topic all of its own. This section is meant to provide only the briefest glimpse into when and why you might consider using EM for your project.

There is also increasing interest in **correlative microscopy**. In correlative microscopy, users analyze the same sample with both optical and electron techniques. This allows users to extract different information from more than one technique from the same specimen. Correlative electron and optical microscopy (often abbreviated **CLEM** for correlative light and electron microscopy) have started to gain some traction but are still not widespread due to the technical difficulties of finding the exact same spot on the sample, aligning images with different pixel sizes, etc. These technical difficulties will probably be solved in the near future so we felt it worthwhile to include a section introducing the EM side of things. For instance, some vendors are starting to offer specialized sample holders that allow sample locations to be indexed so that the same region is imaged using multiple instruments.

In Section 1.4 and sections earlier in this chapter on optical microscopy and super resolution imaging, we discussed the diffraction limit, which is the idea that the resolution of a microscopy technique is limited by the wavelength of the light. In optical microscopy, the shortest wavelength of light usually used is ~400 nm, resulting in diffraction limits of ~200 nm. Highly specialized techniques (e.g., lithography in semiconductor processing – Section 4.20) use shorter wavelength light deep in the UV or even soft x-ray spectral region to achieve higher spatial resolution.

Electron microscopy abandons light altogether and uses beams of high-energy electrons for imaging. Electron energies in many cases can reach 300,000 eV, corresponding to wavelengths of picometers or less. In this way, EMs exceed the resolution limit of techniques that employ optical light and can image features at a significantly finer spatial resolution than optical microscopy can. Since this is a book on optical techniques, this section will provide only a brief overview of the most important aspects of EM so that you can decide whether it is the right technique for you. Please see the references at the end of the chapter for more details on EM.

Since the signal in EMs is usually just the number of electrons making it to the detector, there is no meaningful concept of color. Images can be false colored for clarity, but in general, images are black and white in the EM world. This can actually be helpful sometimes by removing distracting color variations in the sample that you would observe in a light microscope and allowing viewers to focus on high-contrast surface topology, for instance, in an SEM image. The magnification of the SEM can be turned down to be comparable to a light microscope if so desired.

Optically transparent samples are often hard to image in a light microscope, and many of the different optical microscopy techniques discussed previously in this book have been developed to help deal with this specific issue. Samples that are transparent to visible light are not transparent to electrons, so imaging the surface of a transparent sample is sometimes easier in an SEM, assuming that the sample can withstand the vacuum chamber needed for the SEM.

SEMs can be set to have a very high depth of field, meaning that everything appears in focus at the same time, which can also be a very useful feature when examining the surface of a sample with some three-dimensionality to it. Remember, though, electrons do not penetrate very deep into the sample, so only the surface is imaged; anything below the surface will not be detected in an SEM. To address this limitation, samples are often cross-sectioned and examined to reveal their subsurface structures.

Scanning electron microscopes have excellent x–y spatial resolution, but are qualitative in z. If you need to measure the height of something in your sample you will need to view it in cross-section, by either mounting it in the x–y plane or tilting the sample stage in the microscope so that the sample is in the x–y.

Underlying Physical Principles

There are three general classes of electron microscopes: **scanning electron microscopes** (SEMs), **transmission electron microscopes** (TEMs), and **scanning transmission electron microscopes** (STEMs). All three techniques can be described by analogies, denoted here in parentheses, to various types of optical setups described previously. Note that because electrons are charged, lenses in electron microscopes apply magnetic and electric fields to deflect electrons via the Lorentz force rather than using a change in refractive index as optical lenses do. Despite this difference, the general principles of electron lenses can be thought of similarly to optical microscopy and ray diagrams are drawn equivalently, even if the underlying physical mechanism of lensing is somewhat different for light and electrons.

In SEMs, an electron beam (often shortened to e-beam and pronounced "ee beam") with energy ~1–30 keV produced by an electron "gun" (light source – Section 2.2) is collimated at the top of a beam column (optical path or optical axis) using a series of lenses and beam expanders (see Chapter 5) that can focus electrons. The electron beam is then focused to a point. A specialized lens deflects the electron beam to scan it back and forth across the sample. This process is known as "rastering" the beam (scanning mirrors in laser scanning confocal microscopy – Section 4.9). Next, the high-energy electron beam reaches your sample. A variety of processes can occur when high-energy electrons smash into matter. First, electrons may reflect or bounce off of the sample; these electrons that bounce off are known as "back-scatter electrons" (reflected light). Second, the incident high-energy electrons can produce "secondary electrons," with energies <100 eV. Next, the back-scatter and secondary electrons that are produced by the electron beam reaching your sample are detected, and correlated with the position of the beam at that time. Software converts the detector signal into an image. The beam can be focused to an extremely tight spot since the electron wavelengths are tiny, so high spatial resolution (~1 nm) can be achieved using SEM imaging. Since low-energy electrons are highly affected by small variations in the density of material, **secondary electron (SE)** detectors give an excellent idea of the topology of your sample; that is, you can see surface roughness and texture extremely well. **Back-scattered electron (BSE)** detectors yield very high-resolution images and give an idea of the elemental composition of your sample (since the percentage of electrons reflected depends on the atomic number of the elements in your sample), but do not show topology particularly well since high-energy electrons are impervious to small variations in the topology of the sample. High-energy beams penetrate farther into samples. For this reason, low accelerating voltages (≤ 5 keV) are often used for imaging surfaces via secondary electron imaging, whereas higher beam energies are typically used to extract non-quantitative compositional information via back-scatter electron imaging.

Another process that can occur when the electron beam strikes your sample is that incident electrons can excite electronic transitions, which can excite electrons. As the electrons relax, they emit photons that can range from x-rays to visible. Since the x-rays and photons that are emitted from the sample are characteristic of its composition (via specific electronic transitions – see e.g., Section 3.3), x-ray and other photon detectors can also yield spectroscopic information about your sample. The information received by detectors at any given time can be correlated with the beam position in the microscopy software, so SEM can yield detailed composition maps as well as high-resolution images. The two can be overlaid to produce wonderfully informative elemental distribution maps. This technique is known as **energy-dispersive X-ray spectroscopy** (abbreviated in several ways as EDS, EDXS, EDX, and commonly but incorrectly as EDAX, which is actually a brand of detector and software). For optical photons, this process is known as **cathodoluminescence**. Other beam–solid interactions occur too, but are principally associated with highly specialized techniques, so they won't be covered here.

In TEMs and STEMs, the electron gun and focusing apparatus are similar to those in SEMs. However, as the name "transmission" suggests, in TEMs the electrons pass through the sample to reach detectors sitting below the sample. To do this, the electrons must have significantly higher energies (typically 100–300 keV). Some extreme EMs have achieved MeV electron beam energies, but electrons with that much energy vaporize samples quite quickly.

In TEMs, the beam is collimated (i.e., electrons are traveling parallel to the optical axis) as they pass through your sample; in STEMs, the beam is focused and, similar to an SEM, lenses raster the electron beam across the sample. Since electrons are charged, as they pass through your sample they interact with the electron clouds in the atoms of the sample. This adjusts the phase of electrons with respect to each other, causing interference, resulting in high- and low-intensity regions in the beam. The electron beam passes down to a CCD camera (Section 2.3.47) designed to detect electrons. Using TEM, one can view the atomic rows in a thin sample to characterize defects and crystal microstructure.

Additionally, one can operate a TEM in diffraction mode (dark field) and observe diffraction patterns. Since the angle at which electrons are scattered depends strongly on the crystal structure of a sample, diffraction patterns collected in a TEM offer a powerful tool for characterizing crystallography at the nano- and microscales. The diffraction patterns are actually *k*-space (Fourier optics).

Scanning transmission electron microscopes are relatives of TEMs, except they combine the focused, rastering beam of an SEM with the electron transmission characteristic of a TEM. Therefore, they can also achieve extremely high-resolution images, but can combine this high resolution with location-dependent elemental mapping and diffraction.

What Scientific Questions Can Be Asked?

SEM/EDS can tell you:

- What does your sample look like to ~1 nm resolution?
- What is the topology of your sample? Note, however, that the topology is not quantitative unless you are looking at the sample in cross-section.
- What is the elemental composition of the top ~1000 nm of your sample?
- How does the elemental variation of the surface of your sample vary, with ~250 nm resolution?

TEM/STEM can tell you:

- What is the atomic microstructure of a small slice of your sample?
- What types of crystal structure and atomic defects does your sample have in that tiny slice?
- What is the microscopic composition of your sample?
- How does the elemental variation of your sample vary, with approximately sub-nanometer resolution?

Strengths and Limitations

Electron microscopy offers extraordinarily high spatial resolution; SEM can achieve nanometer-scale imaging and TEM/STEM can resolve the atomic crystal structure of your sample; SEM also offers a high **depth of field** (i.e., length in optical axis in which your sample can be in focus) relative to optical techniques.

However, EM is limited by a small field of view, especially for TEM/STEM, and relatively onerous sample preparation procedures since the samples need to reside in high vacuum. (Exceptions in modern SEMs are beginning to enable imaging for non-vacuum compatible samples, but these are not yet standard at most institutions.) Furthermore, labels for biological samples are limited compared to fluorescence microscopy. Lastly, SEM can only image the surfaces of samples. One must perform some sort of (serial) sectioning, cross-sectional imaging, or advanced tomography technique in order to understand the insides of a sample.

What Samples Are Appropriate?

Since electrons have charge and mass, they deflect off of the other electrons in atoms. Therefore, the entire electron microscope setup is held in high or ultra-high vacuum (10^{-6} to 10^{-10} Torr) to minimize deflections of the electrons from anything but your sample. So, samples must be compatible with a high-vacuum environment. That means that samples should have low vapor pressures so they don't evaporate once you put them into the vacuum. Examples of materials that have low vapor pressures are most metals, ceramics/dielectrics, and semiconductors. Materials to avoid are organics such as adhesives from tape, biological samples, foods, etc.

For SEMs, samples must also be somewhat conductive so they don't "charge up," meaning that electrons stick to the surface of an insulating sample to generate a negatively charged surface. This charged surface deflects the other electrons from the beam and makes imaging the sample quite difficult. One can sputter coat thin (~1–2 nm) layers of gold or other metals onto most samples to minimize the impact of surface charge-up. Properly grounding samples using metal clips and copper tape is also helpful to prevent charging.

That said, a spectrum of specialized techniques has enabled electron microscopy analysis of biological, food, or other organic/wet samples. In particular, one can dehydrate or otherwise "fix" samples to make them compatible with the high vacuum inside the microscopes; stain or mineralize organics; or image samples at ultracold ("cryo") temperatures, keeping the glassy ice in a state where it will not sublimate into the vacuum system or crystallize, which will often destroy samples. Some new SEMs even have "environmental" modes, where one can image samples at ~1 Torr of water vapor instead of high vacuum.

Samples also need to fit inside the vacuum chamber. Most modern SEMs can accommodate samples up ~10 cm in diameter and ~1 cm in height. For TEMs and STEMs, the sample must be extremely thin (~<200 nm thick) so that the electrons can pass through, and must fit inside the sample holder (~2 mm diameter, depending on the type of microscope). Specialized techniques such as cross-sectioning via microtome, polishing, and focused ion beam (FIB) lift-out can extract extremely thin samples (aka "lamellae") from larger samples that are appropriate for TEM and STEM.

Common Pitfalls

Electron microscopes are extremely complicated (Figure 4.71) and expensive (~$500,000 to $1,000,000 for SEMs; $1,000,000 to $5,000,000 for TEMs and STEMs). Therefore, it takes significant training (about five hours for an SEM and many, many days for a TEM or STEM) to learn to use EMs. Do not assume you will extract useful information from your sample the first time you use the

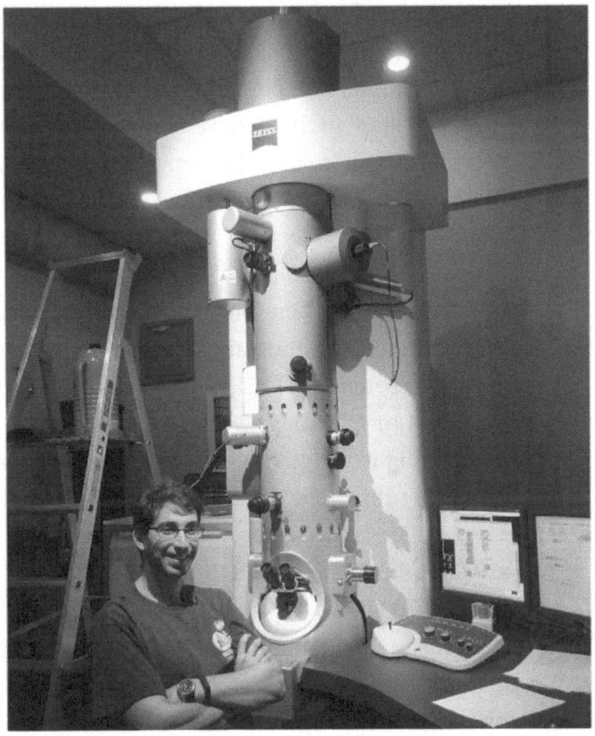

Figure 4.71

One of the authors in front of an aberration corrected TEM at Harvard. Hopefully this picture gives you some idea of the complexity and scale of these microscopes!

microscope. Also, given their cost and complexity, EMs typically have dedicated maintenance and training staff and reside in specialized facilities that minimize vibration and other sorts of noise, so they are only really available at large companies, universities, and dedicated scientific/engineering analytical facilities.

That said, many companies have recently been going out of their way to reduce the cost and complexity of their EMs. For instance, trained operators can acquire useful images within about ten minutes of loading a sample into a modern SEM (from companies such as FEI and Zeiss) that feature fast pump-down times, easy alignment procedures, and powerful but intuitive software packages. Other companies (e.g., JEOL and Hitachi) offer desktop SEMs that do not have the resolution of larger microscopes but are comparatively quite simple to use and inexpensive. One company, Voxa, even offers miniature EMs, with a correspondingly low price tag. Amazingly, these microscopes from Voxa can be operated in SEM, TEM, and EDS mode, and are the size of a carry-on suitcase.

One other word of caution: the physics of electron beam/matter interaction are quite complicated, and are mostly beyond the scope of this book. Consult with your institution's EM expert on exactly how to interpret your data, and which specific technique (both for imaging and for sample preparation) is appropriate for you.

For example, one common misconception concerns the resolution of an EDS elemental map. In SEM, the energies of the electron beam are high. Therefore, the electrons penetrate into the sample to what is known as an "interaction volume." This is a region in the sample that can be ~1000 nm from the surface of the sample, depending on the energy of the electron beam. The interaction volume also extends out in the plane of the surface from the point at which the electron beam hits the sample. Therefore, x-rays can be generated from the entire interaction volume, since electrons are spreading out underneath the surface of the sample. So, the resolution of EDS mapping is significantly limited compared to the resolution of the images, and even though you may set the pixel size to 2 nm, don't assume that the x-ray information you collect from each pixel strictly emanates from areas equal to the step size of the beam raster.

Sample Data and Data Interpretation

Figure 4.72 shows a high-resolution TEM (HRTEM) image of gold nanoparticles suspended on an amorphous carbon film. The gold particles are the dark objects. From this image, one can clearly see the atomic structure of the particles. The regular patterns that can be seen correspond to electron diffraction from the alternating regions of electron density that correspond to the crystal lattice in the particles as thc clcctrons pass through the sample. Irregularities in the crystal structure correspond to stacking faults and other microstructural defects that can be correlated with properties such as surface plasmon resonance frequency or catalytic activity. Such information can also be used to adjust the synthesis of the particles to tune the crystal structure. The particle diameters can accurately be measured to ~20 nm. Moreover, one can measure the angles between surface facets, which can inform users about the identity and crystal structure of the surfaces on the particles.

Note that TEM images yield two-dimensional projections of three-dimensional objects. So, it is important not to infer much about the 3D structure of these particles from this image alone. TEM tomography techniques would need to be applied for 3D information. In TEM tomography, the sample would be rotated and images would be acquired at as many angles as possible. The information from the images at different angles would then be reconstructed by software into a 3D view of the sample.

Another important point is that since TEM has such a limited field of view, statistics are hard and time-consuming to collect. Images of many other particles would need to be aggregated to form an accurate representation of the population of nanoparticles as a whole.

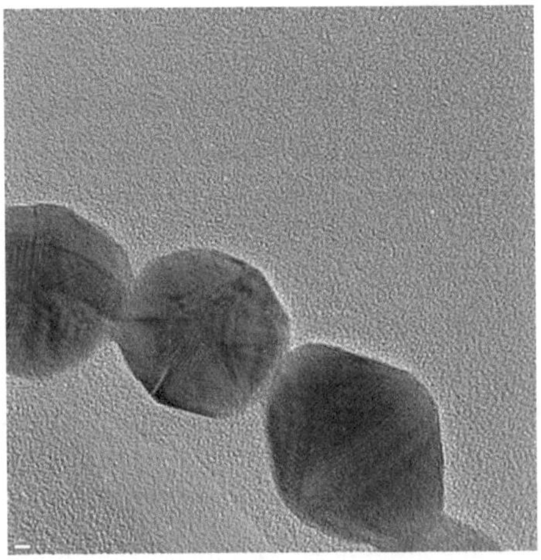

Figure 4.72

TEM image of gold nanoparticles suspended on a carbon film. Lattice fringes corresponding to columns of atoms in the gold particles are clearly visible. Scale bar, 2 nm.

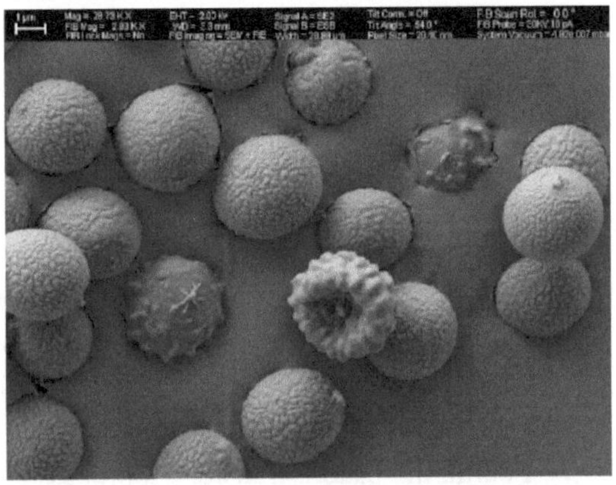

Figure 4.73

Cryo-SEM image of bacteria cultured from a swab of a lab keyboard at Harvard. Gross!

Figure 4.73 shows a cryo-SEM image of bacteria cultivated from a swab of a lab keyboard at Harvard. In this case, cryo-SEM was used to visualize the structure of the bacteria without having to perform complicated dehydration steps. This image was recorded using the instrument's secondary electron detector and an acceleration voltage (electron energy) of 2000 V (2000 eV), to capture the surface topology of the bacteria that are embedded in ice. One can clearly visualize the surface structure of the bacteria and their size, but from this image one should not infer anything about the interior structure of the bacteria, or the atomic compositions of the bacteria or their surfaces.

4.20 Photolithography: A Brief Overview

Photolithography is a widely used technique in the semiconductor industry that is used to pattern small structures on a substrate, typically a silicon wafer. This process is essential to the fabrication of transistors on computer and cell phone chips, and other microelectronics such as gyroscopes and accelerometers for cell phones, random access memory (RAM), microfluidic diagnostic devices for biomedicine, small radio antennas and RFIDs for ID cards, flat panel televisions, and solar cells.

Since photolithography relies on some optical principles that can be understood by analogy to microscopy, we will give a brief overview of the technique here.

A basic photolithography process is shown in Figure 4.74. First, an organic compound known as a photoresist that is sensitive to blue and UV light is coated onto a substrate with controlled thickness, usually by spin or spray coating. The photoresist can operate in one of two modalities: a "positive" photoresist comprises extended polymers that undergo electronic transitions when illuminated by UV light, causing them to break apart into subcomponents that become soluble. "Negative" photoresists respond to UV light by polymerizing and become insoluble. The process in Figure 4.74 depicts a positive tone photoresist.

Next, a photomask is placed in close contact with the photoresist-coated substrate. Photomasks typically consist of glass plates with chromium patterns on them representing some part of the geometry of the device to be fabricated. The glass regions are transparent to UV light, and allow the light to expose the photoresist only in selected areas. Chromium blocks the UV light, so the resist under the chromium regions remains unexposed.

Next, the photoresist is developed in a solvent that is specifically designed to remove only the exposed (unexposed) portions of the positive (negative) resist.

Fourth, in "pattern transfer," a process is performed to alter the substrate. This process could be an etch, to remove material selectively in areas that are not covered by the photoresist, or a deposition, to add materials selectively in that region. Finally, the photoresist is removed.

Ultraviolet light is used in typical photolithography systems because it has a shorter wavelength compared to visible or IR light. Due to the diffraction limit of light, shorter wavelengths allow for smaller resolutions. Specifically, the minimum feature size achievable with contact lithography is denoted:

$$CD_{min} = k\sqrt{\lambda d} \tag{4.4}$$

where CD stands for critical dimension, k is an experimental constant between 1 and 2 that is associated with the lithography equipment, λ is the wavelength of

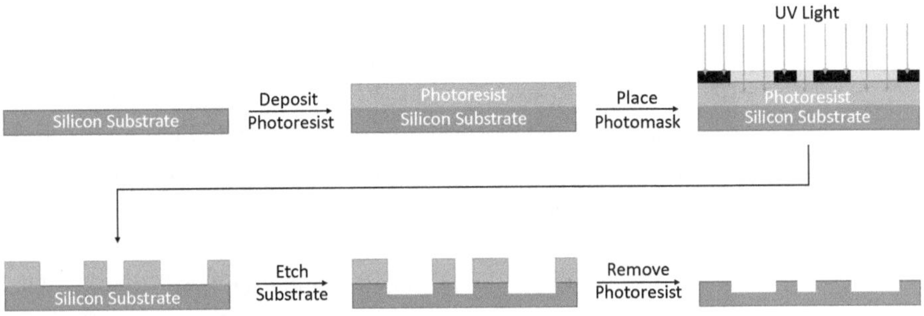

Figure 4.74

A basic photolithography process. First, a silicon substrate is coated with a thin film of photoresist. Then, the glass contact mask is placed on the wafer. The mask and wafer are exposed to UV light, and the mask is removed. Next, the wafer is developed in solvent meant to remove the exposed areas of photoresist. Next, a process is performed on the Si wafer, which only affects the wafer in the areas exposed by the openings in the photoresist. Finally, the photoresist is stripped away.

the light, and d is the thickness of the photoresist. Therefore, to achieve better resolution, one can use thinner photoresist or lower wavelengths (higher energies) of light. However, in practical reality, defects in the photoresist or the mask typically limit the resolution of contact lithography to 1–2 μm feature sizes and pitches (i.e., spacing between features).

To overcome these limitations, the semiconductor industry has developed advanced lithography techniques such as projection or stepper lithography, extreme UV (EUV) lithography, immersion lithography (whose working principle is analogous to the resolution gained by using an immersion lens in microscopy – Section 4.1), and electron-beam lithography. These techniques each have advantages and disadvantages, and are consequently deployed in different settings. For instance, electron-beam lithography is common in university labs because it can routinely achieve extremely fine resolutions <100 nm (Section 4.19), but is slow and expensive to run, so it is inappropriate for industrial applications. Projection lithography is faster, and can beat the resolution of contact lithography. In essence, projection lithography uses a microscope to expose a photoresist-coated wafer to a small beam of light. Therefore, similar to optical microscopy (Sections 1.4 and 4.1), the resolution achievable with projection lithography is:

$$CD_{min} = \frac{k\lambda}{NA} \tag{4.5}$$

where NA is the numerical aperture of the projection optics. Again in this case, analogously to microscopy, smaller wavelengths yield finer resolutions. Therefore, next-generation optical lithography is projected to use even lower wavelengths of light, all the way into the EUV. Extreme UV lithography is designed with high-volume industrial production in mind, so tool costs begin at several tens, if not hundreds, of millions of dollars. The next generation of EUV tools is projected to use 13.5 nm (91.8 eV) light; the choice of wavelength is primarily dependent on the availability of stable light sources; in this case a tin plasma serves as the gain medium in a laser cavity. A previous generation of lithography tools operating at 193 nm (6.4 eV) employed argon-fluorine excimer lasers (Section 2.2.3).

Photolithography is complex, and each fabrication facility typically has specialized photoresists and specific processes developed for its tools. Often times, new process development is required for different applications and substrates. We recommend speaking with your fabrication facility staff to better understand specific lithography capabilities at your institution.

4.21 Further Reading on Microscopy Techniques

Fundamentals of Light Microscopy and Electronic Imaging by Douglas Murphy and Michael Davidson is an excellent introductory text to different optical microscopy techniques.

Imaging: A Laboratory Manual by Rafael Yuste is a useful manual focused entirely on imaging of biological samples.

Immunocytochemistry: A Practical Guide for Biomedical Research by Richard Burry is a very useful guide to antibody fluorescent labeling techniques that are used in sample preparation of biological samples for many of the fluorescent microscopy techniques.

Culture of Animal Cells by Ian Freshney is a practical guide to cell culturing techniques. It is an excellent resource for those new to keeping cells happy and healthy in a lab.

Live Cell Imaging: A Laboratory Manual by Goldman, Swedlow, and Spector is a higher-level book with contributed sections from a variety of authors. It is useful as you get deeper into microscopy techniques for live cell applications.

Atomic Force Microscopy by Peter Eaton and Paul West is a good introduction to AFM.

Physical Principles of Electron Microscopy: An Introduction to TEM, SEM, and AEM by Ray F. Egerton is a comprehensive overview of EM techniques.

Transmission Electron Microscopy by David B. Williams and Barry C. Carter is the go-to four-volume text for expert TEM users.

5 Notes on How to Design and Build Optical Setups in the Lab

When you start to build a new optical setup, there are some points worth keeping in mind to make your life easier as you use the setup to collect data. Some of these tips have been embedded in the other sections of the book, but they are repeated here for those who have not read the book cover to cover.

If you get the opportunity to start building an optics lab from an empty room, think about accessibility to all sides of the optics table. Being able to walk all the way around the optics setup often makes life much better as opposed to having to carefully reach over a beam path to adjust something on the far side of the table. Arms and clothing seem to invariably dip into the beam path you are reaching over, which can be dangerous with high-powered lasers. Try to resist the urge to put the table up against a wall, since this placement limits access to one side of the table.

Next, have an overhead rack built to hold power supplies, control computers, etc. The overhead rack should be mounted to the ceiling, independent of the optical table. This will mechanically decouple all these sources of vibration from your sensitive setup. Generally, it is better to position the rack higher so that tall people don't bump their heads every day. (From personal experience, it is easier to store a small step stool under the optics table for shorter lab members than to have the tall lab members constantly banging their heads.) You will want a number of electrical outlets (probably in the range of 15–20) mounted on the overhead rack. Plugging things into the wall outlets invariably leads to tripping hazards in the dark.

Having a workstation for the computer screen, keyboard, and mouse mechanically attached to the overhead rack is also a good idea. You don't want vibrations from your typing showing up in your data. Avoid seated computer workstations that put your eyes at the level of the optical table. Standing workstations or workstations with tall stools that keep your head well above the plane of the laser are much safer.

Make sure that you have proper laser curtains, "Laser in Use" signs, and laser goggles. The laser goggles should live in a cubby by the door to the lab. If you have multiple wavelengths of lasers in the lab, keep the goggles sorted so that lab members don't accidentally use the wrong laser goggles. Ideally the "Laser in Use" sign switch should also be by the door so it can be turned on as a habit when you enter the lab to start experiments. The "Laser in Use" sign should be bright enough to get people's attention and positioned at a height that people will see. Often, these signs are ceiling-mounted and people walk right under them, unaware of their message. Also, please remember to actually turn the signs off when the laser is off so that other lab members don't learn to ignore the sign.

Your optics lab should have a source of compressed dry air, usually a pipe on the wall, to float the optics table. Think about how to route the airline so that it is not a tripping hazard in the dark. If you can, route it up and overhead, and then down to the table legs. Check that the optics table is truly floating. To do so, gently push on the corner of the table. The whole table top should move. As you release the table, you should hear a hiss of the legs refilling with air.

Check around the lab for reflective surfaces. For instance, glass-fronted cabinets look nice, but are a hazard. Beware of stray beams reflecting around the lab in an uncontrolled manner.

In an American optics lab, you will invariably end up with both a set of imperial screws and wrenches and a set of metric screws and wrenches. Figure out a way to clearly label them. Nothing ruins a ¼-20 (imperial) threaded hole faster than forcing an M6 (metric) screw into it. If you are unsure which system screw or wrench to use (metric or imperial), consider where the equipment was manufactured. For instance, Newport and Coherent are American laser companies out of California so they use imperial screws and wrenches. The four research-grade microscope manufacturers are from Germany (Zeiss and Leica) or Japan (Nikon and Olympus), so they will use metric screws and wrenches. Brute force is not the answer if a screw doesn't seem to be fitting.

Resist the urge to just start putting optical components on the table. There is always pressure to start collecting data as soon as possible, but some upfront planning will make your life much better in the long run. Once you have built the setup, you will be loath to tear it apart to make it more user-friendly or ergonomic. To begin assembly of the setup, sit down with a cup of your favorite coffee and draw out the optical setup on paper. Consider questions that make the setup easy to use, such as which optical components will you need to adjust regularly? Locate those components of the setup near the edge of the optical table so they can easily be reached.

When you are ready to start laying out the optical setup, take a moment to draw out the path on the optical table. Sharpie markers write well on the stainless-steel optical tables and the lines can easily be removed with acetone and a paper towel. Use the lines of holes on the optical table as guides to help you make sure that your setup is using straight lines and right angles. Optical setups with straight lines and right angles help you to preserve the polarization of the laser light.

Always use two mirrors to steer a laser beam. One mirror only allows you to change the angle of the beam, while two mirrors allow you to translate the beam path. Place these two mirrors far enough apart (~12" or greater if possible) such that a small adjustment to the mirror adjustment knobs will produce a noticeable effect on the beam position. This large spacing may seem like unnecessary extra optical components and space on the table, but it is needed for good control of the laser beam.

Remember that the optical surface of the mirror is what needs to be in line, not the post. Many mirror mounts actually sit significantly in front of the post and a common error is positioning the optical posts in the beam path, when in fact the mirror surfaces really need to be in the path. Double-check that dielectric mirrors are positioned properly in the setup, since dielectric mirrors are designed for the laser beam to be incident at a specific angle. If the laser beam is not at the correct angle of incidence, the dielectric mirror may not be reflective.

Set up the beam path with mirrors first (Section 5.1.2); then insert any other optical elements like lenses, polarizers, half-wave plates, etc. A crooked optical element (lenses especially; Section 5.3) can alter the path of the beam up, down, left, or right, which can cause problems further down the optical path.

Always keep the beam in the plane of the optical table. This practice keeps the right angles at right angles and it also keeps the laser away from eye level.

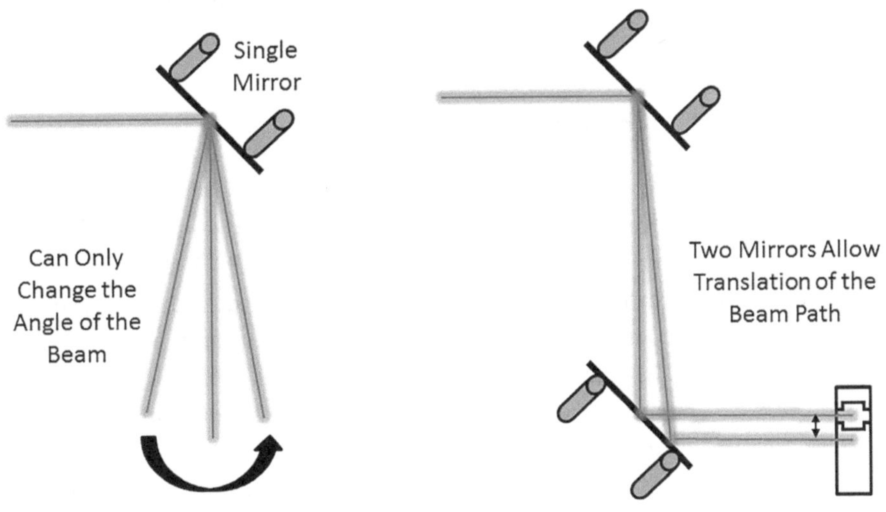

Figure 5.1

Always use at least two mirrors (right) to guide a laser beam through an optical setup. This allows you to translate the path of the laser beam. A single mirror (left) only allows you to change the angle of the beam path.

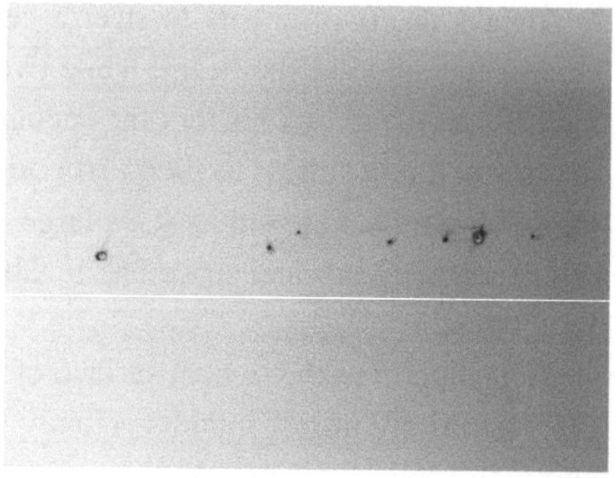

Figure 5.2

Laser burn marks on the wall from when a femtosecond laser got away from someone during the adjustment of a mirror for a new optical setup. Wear your laser safety goggles!

When you are ready to start sending the laser beam through your optical setup for the first time, keep the laser power as low as possible but high enough that you can still see it on a viewing card. Minimize the number of people in the room while adjusting the laser beam. It is relatively easy to send the laser in an unintended direction while you are adjusting mirrors. Most femtosecond laser labs have laser burn marks on the walls where the beam has gotten away from someone during the setup of a new optical path (Figure 5.2). Move through the setup one mirror at a time, keeping track of where the laser beam is going. Try to keep a beam block in place in front of the next component in the setup.

5.1 Cleaning Optics

Wear gloves when handling optics to keep fingerprints off the optical components. Keep the optics clean with an air puffer. (A puffer here is a rubber bulb that puffs air when you squeeze it; see Figure 5.3.) We do not recommend compressed air cans that are often used for cleaning dust out of keyboards and personal computers. The compressed cans of air can squirt out a liquid that sometimes leaves a residue on the optics when the can is held at a non-vertical angle, so we recommend avoiding them. Trying to blow the dust off with your mouth will often result in spit residue on your optics.

Always start by using some air to blow the dust off of an optical component that needs to be cleaned. This will hopefully remove any dust or grit that might scratch your optic when you wipe it. For optics that need further cleaning, fold up a piece of lens cleaning tissue (Kimwipes are *not* lens cleaning tissue!!!) into a small pad while wearing gloves. (Your fingers are oily and this oil can deposit on the optics

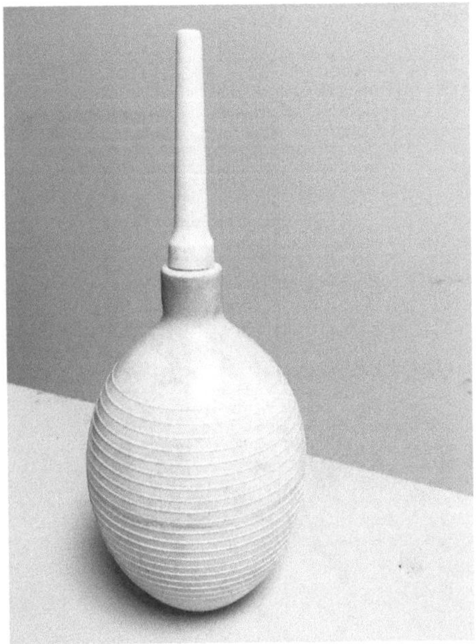

Figure 5.3

Air puffer used to clean optics.

and disrupt their function. Additionally, in high-power laser systems, oil from fingers can absorb light, heat up, and melt the otherwise transparent optic.) Hold the lens cleaning tissue pad in a pair of locking forceps. (We prefer curved forceps, but straight forceps will work too.) Add a couple drops of mild solvent. Microscopists swear by Sparkle Glass Cleaner (the purple solution in the house cleaning aisle) for cleaning microscope objectives. Spectroscopists tend to swear by isopropyl alcohol or methanol. Shake the excess solvent off of the lens cleaning tissue. Make one gentle wipe across the optic. Toss the used lens cleaning tissue in the trash. If the optic needs more cleaning, start again with a fresh lens cleaning tissue. Do not scrub the optic. Do not use each lens cleaning tissue for more than one wipe! If the lens cleaning tissue picked up any particles, these could scratch the optic if you use it again.

5.2 Walking the Beam

Once you have decided what your optical path needs to be, you will need to guide your laser beam through the path. First, you will need to build a laser beam height target. Mount a viewing card that allows you to see the beam through your laser

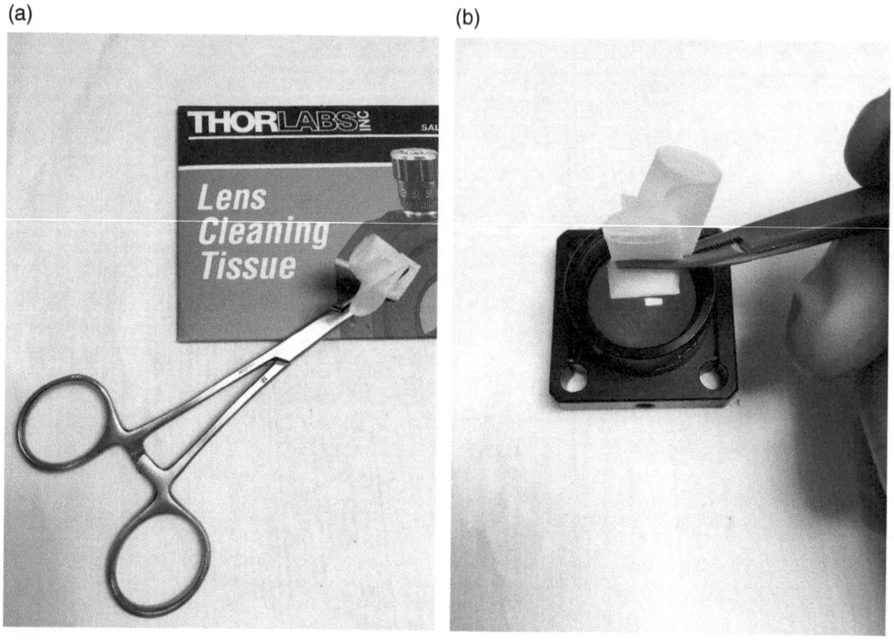

Figure 5.4

Cleaning optics using lens cleaning tissue. (a) Using forceps to secure a piece of lens cleaning tissue. (b) Cleaning the lens.

goggles on a post (or a ruler) and place the post in a post holder. Clearly mark the point on the card that you want to hit. Fluorescently colored index cards or sticky notes from an office supply store often work as viewing cards. Since they are cheap, you can cut them to size and mark beam positions on them without feeling bad.

Remember that two points define a line and that laser beams should travel through a straight line parallel to the optical table. Pick two positions in the proposed beam path. Ideally the two positions should be several feet apart, as the farther apart the positions are, the more sensitive to small misalignment this process will be.

Coarsely adjust the mirrors to get the laser beam traveling in the proper direction. Tighten the posts of the mirror mounts with a hex wrench using the locking thumb screw (Section 2.3.3). For fine adjustments, use the adjustment knobs on the first mirror to adjust the position of the laser on to the target at the closer of the two positions (Position 1 in Figure 5.5). Next, move the target to the second position in the beam path (Position 2 in Figure 5.5). Use the second mirror to adjust the laser beam position onto the target.

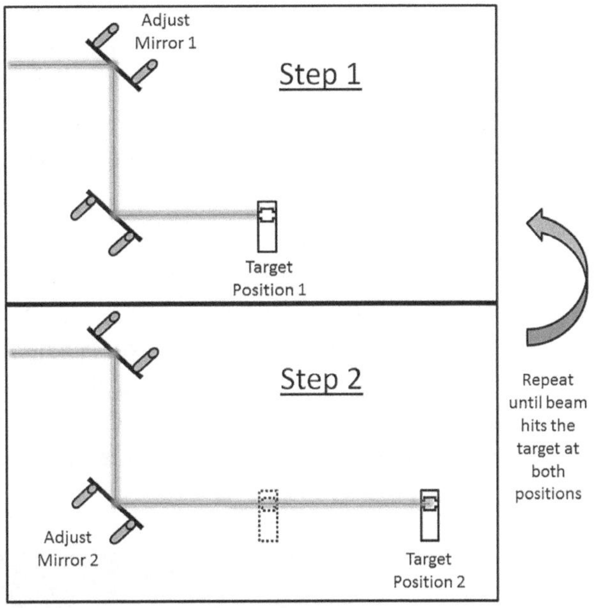

Figure 5.5

Walking the laser beam. When walking the beam, use mirror 1 to adjust the beam position when the target is in position 1 and use mirror 2 to adjust the beam position on the target when it is in position 2.

Iterate the above process several times, and be sure to switch which mirror you are adjusting to translate the beam into the desired optical path.

Finally, mark the desired beam path with adjustable irises so that if something gets bumped, you can quickly recover the desired optical beam path. The adjustable irises act as the laser beam height targets when you close them.

5.3 Inserting a Lens or Other Optical Element

While mirrors are usually mounted in kinematic mounts to allow steering and translation of the laser beam, lenses and other optical elements are often mounted in fixed mounts with no angle adjustment knobs. This does not mean they cannot steer the beam when inserted into the optical path. In fact, care must be taken to ensure that you do not accidentally steer the beam when you insert a lens or other optical element in a fixed mount. The laser beam should be sent through the desired optical path first with only the steering mirrors; then insert the lenses into the optical path one at a time.

When inserting a lens, start by checking the height of the lens. You should be sending the laser beam through the middle of the lens. Use the beam height targets after the lens to check that you are not steering the beam up or down with the insertion of the lens. The beam should still be in the same plane parallel to the optical table before and after you insert the lens. Next, check that the lens is perpendicular to the beam. Use the retroreflection from the lens to check the left–right rotation of the lens. The retroreflection is the back-reflected beam from the lens. When the lens is perpendicular to the laser beam, the retroreflection will be exactly aligned with the incoming laser beam.

To look for the retroreflection, use a laser viewing card and place it next to the incident laser beam. Rotate the lens in its post holder. A collar (Section 2.3.5) on the post will keep the lens at the right height. A reflected laser spot should appear on the viewing card. Carefully rotate the lens back until the retroreflection disappears into the incoming laser beam. Check on the other side of the incoming laser beam with the viewing card to ensure that you have not rotated the lens too far.

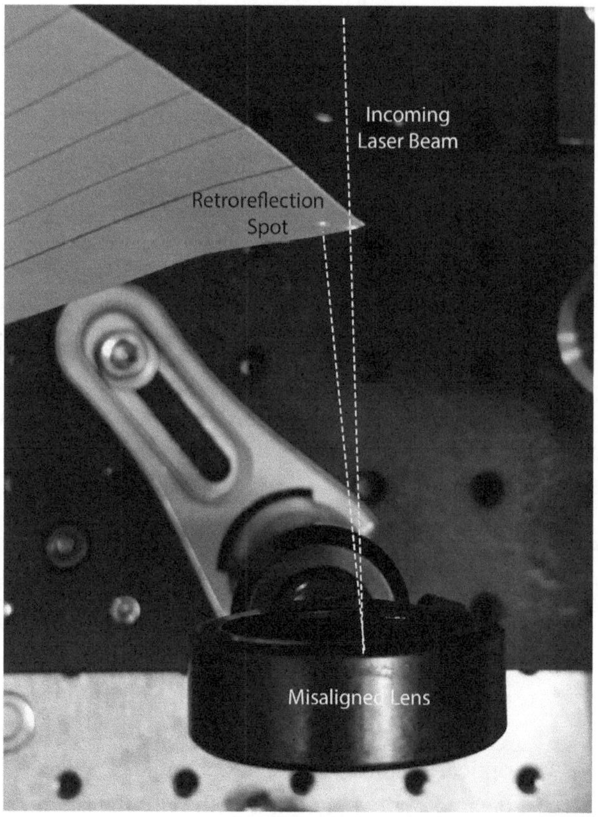

Figure 5.6

Retroreflection from a lens. Checking that the retroreflection is aligned with the incoming laser beam allows you to check that the lens is perpendicular to the incoming laser beam.

Check above and below the incoming laser beam for retroreflections also. If there is a retroreflection above or below the incoming laser beam, your lens is crooked in the mount and it will need to be re-mounted. If re-mounting does not help, check that your incoming laser beam is parallel to the plane of the optical table by placing the beam height target at several positions in the optical path. If the incoming laser beam is not parallel to the optics table, walk the beam (Section 5.2) until it is parallel to the table.

Be careful that you do not have a retroreflection going straight back along the incoming laser beam path into your laser though. Retroreflections that make it all the way back into your laser cavity can destabilize or even damage your laser. Faraday isolators are often set up at the output of lasers to ensure that retroreflections do not make it back into your laser cavity. Check your laser's manufacturer's specifications to be sure.

5.4 Plumb Line

If your beam path ever needs to leave the plane of the table, it is beneficial to ensure that it travels in a straight, vertical line. Plumb lines are simple tools to help with ensuring that vertical lines are truly vertical.

A plumb line is a piece of string with a weight on the end. A couple washers on the end of a string will work as the weight.

To use the plumb line, block the laser beam. Use the plumb line to determine exactly where on the ceiling the laser beam should hit if it is traveling perfectly vertically. Mark the spot on the ceiling with a viewing card. Turn the laser back on and adjust the mirror that is directing the laser beam upwards so that the beam hits the target on the viewing card that was determined as perfectly vertical by the plumb line. Be very careful not to bend over the laser beam that is traveling vertically. Laser goggles are not usually designed for beams coming vertically toward you. Beware the laser beam coming up under the goggles by accident.

When coupling a laser beam into an inverted microscope, first turn off the laser and remove the microscope objective from the beam path. Use a plumb line to determine where a beam coming through the center of the microscope objective port would hit the ceiling. Mark the spot on the ceiling with a laser viewing card. Turn on the laser and walk the beam using the center of the microscope objective port and the target on the ceiling as your two targets. Finally, reinsert the microscope objective into the beam path. Often it is desirable for the laser beam to fill the back focal plane of your microscope objective. (TIRF microscopy [Section 4.8] is the one notable exception to this.) In modern microscope objectives, the back focal plane is designed to be the last piece of glass on the back of the

(a) (b)

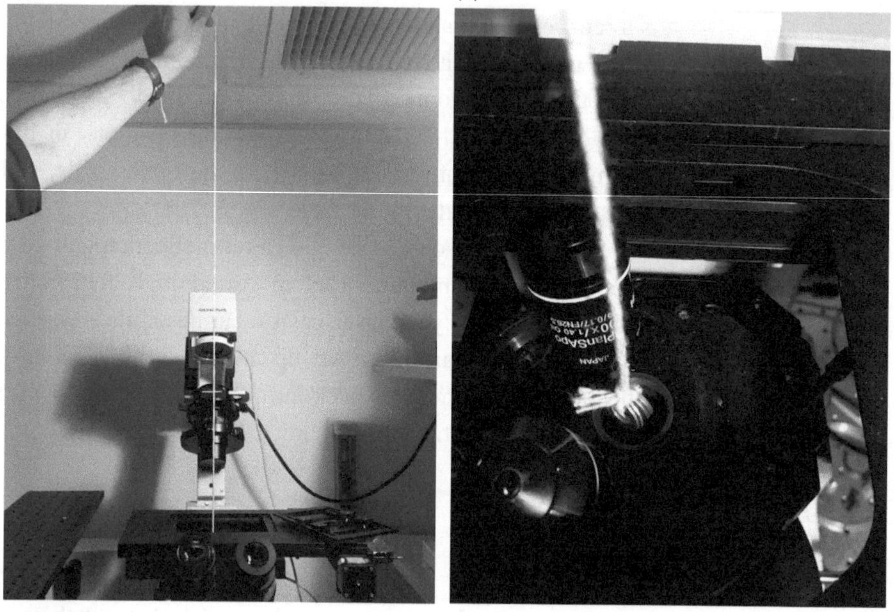

Figure 5.7

Plumb line being used to find true vertical from an empty microscope objective port.

objective. If the diameter of the laser beam at the microscope objective port on the nosepiece is not large enough to fill the back focal plane of the objective, you should add a beam expander (Section 5.9) to your optical path.

5.5 Periscopes

You will periodically find it necessary to change the height of the beam path in your setup. To do so, use two mirrors arranged like a periscope. To set up a periscope, first block the laser beam. Position the lower mirror in the beam path of the laser, at approximately 45 degrees to vertical and the horizontal laser beam line. Use a plumb line as described in Section 5.4 to determine where a perfectly vertical optical path is. Turn on the laser and adjust the bottom mirror of the periscope such that the laser spot tracks the vertical optical path. Turn off the laser beam. Insert the upper mirror of the periscope. The upper mirror should send the beam in the same direction on the table that the beam was originally traveling only at a different height from the table.

Figure 5.8

Periscopes used to change the height of the laser beam. (a) Straight periscope which only changes the height of the beam and (b) a folded periscope which changes the height and flips the polarization of the laser beam. For instance, if the beam starts as horizontally polarized, it will be vertically polarized after a folded periscope.

If you position the upper mirror such that the beam path also changes by 90 degrees then you have made a folded periscope. Be aware that a folded periscope will change the polarization of the laser beam. Sometimes this is desirable, such as when working in the IR region where there are not great broadband half-wave plates available.

5.6 Optical Delay Lines

When using multiple pulsed lasers in experiments such as a pump–probe experiment, stimulated Raman spectroscopy, or STED microscope, controlling the timing between the two pulses is essential. The typical setup for controlling the time delay between the pulses is an optical delay line, shown in Figure 5.9.

Using a piece of string and your fingers to measure the length of an optical path is a surprisingly accurate first approximation. First, use your fingers to pinch the string at each reflection and then compare the total lengths of the two optical paths by measuring the two pieces of string against each other. Adjust the path lengths until the string comes out the same length for both optical paths. Then, start the fine

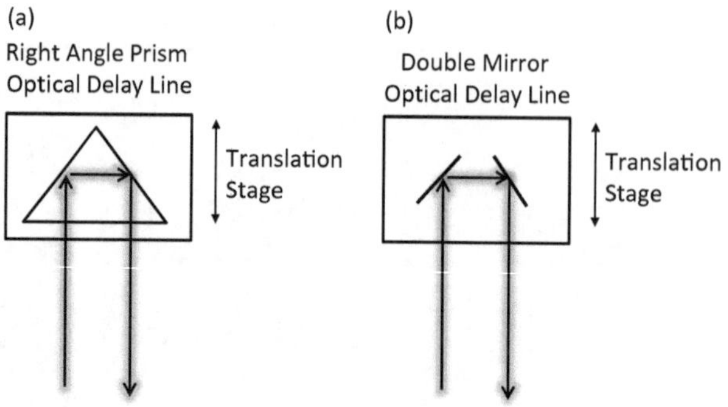

Figure 5.9

Optical delay line ray diagrams.

adjustments using the sum frequency signal from a *z*-cut quartz reference sample (Section 3.8).

A right-angle prism can be used as a delay line mirror, since the angle of incidence is more than the critical angle for total internal reflectance. The single prism can be mounted on a translation stage. Cube mirrors or retroreflectors can also be used as they send the incoming light back in a parallel path. You could also use two individual mirrors on kinematic mounts on a translation stage.

The biggest challenge is usually setting up the translation stage parallel to the beam path such that the beam direction does not change as the delay line is translated.

5.7 Prism Pulse Compressors

The shorter the laser pulse, the wider the spectral bandwidth. A 100 fs laser pulse usually has >50 nm bandwidth. That means that even though people say that they have an 800 nm excitation beam, that is really just the center wavelength. They are really using a pulse with 775–825 nm excitation light, for example. The different wavelengths will have different dispersion as they propagate through the optical setup. This is due to the fact that the index of refraction of glass has a slight wavelength dependence, which is no longer negligible when working with ultrafast laser pulses. While the speed of light is constant in a vacuum, the wavelength dependence of the index of refraction means that the shorter wavelength light will travel faster through a piece of glass compared to the longer wavelength light in the pulse. This will lead to pulse broadening. For example, if you started with a 100 fs pulse at your laser output, it may be 300 fs by the time that it reaches the sample.

We can correct this pulse broadening by using a prism pulse compressor. Here is an analogy to explain how this works. Imagine a tour group at a museum. They all start together but as the tour group travels through the museum it starts to spread with the slower walkers falling farther behind. During the tour, the group leader may pause to let the slowest walkers catch up with the rest of the group. In a similar way, the longer wavelengths in the pulse will fall behind, making the entire pulse longer. While pausing the laser pulse is not feasible, the prism pair pulse compressors give the slowest wavelengths a shortcut so that they can catch up with the shorter, faster wavelengths in the pulse.

To give the longer wavelengths a shorter optical path, we use a prism. A prism will spread out light as a function of wavelength. (A rainbow emerging from a prism is the common example.) By sending the laser pulse through one prism to spread the light out by wavelength, a prism pulse compressor gives the longer wavelengths a shorter optical path and then puts all the light back together with a second prism. This allows the slower wavelengths to catch up with the fast wavelengths.

The term **chirp** in optics refers to the difference in arrival time between the reddest photon and the bluest photon in a laser pulse. A **negative chirp** means that the longer wavelength light is arriving before the shorter wavelength light. A **positive chirp** means that the shorter wavelength light is arriving before the longer wavelength light. Depending on what else is in your optical system, it may be desirable to actually give the reddest wavelengths of light a head start so that all the photons arrive at the sample at the same time.

To build a pulse compressor you will need two equilateral prisms of appropriate material for the wavelength of light that you are using. You will need to be able to rotate and translate the prisms. Start by sending the laser beam into the first prism and rotating the prism to the minimum deviation angle. If you watch the laser beam on a viewing card as you rotate the prism, the beam will move horizontally with the prism rotation until a certain angle when it will stop moving. If you continue to rotate the prism, the laser beam will actually start to move back in the other direction. This is referred to as the **minimum deviation angle**. Set up the second prism ~30 cm away and position it so that it catches the entire beam that is spread out by the first prism. By adjusting the second prism also to the minimum deviation angle you will overlap all the wavelengths spatially again. You should have a tight laser beam again after the second prism. Position a mirror to send the light back through the prism pair. After the first prism, position a mirror as close to the input beam as possible without impinging on the input beam. Use the mirror after the second prism to steer the return beam path to hit the edge of this mirror and guide the compressed pulse out of the prism compressor out into the rest of the optical setup (see Figure 5.10). Unlike lenses, using the edge of a mirror won't cause issues.

By translating the first prism in the laser beam path so that the input beam hits closer to the tip or deeper toward the base of the prism, you can control the amount of dispersion introduced to the beam, which controls chirp or compression of the

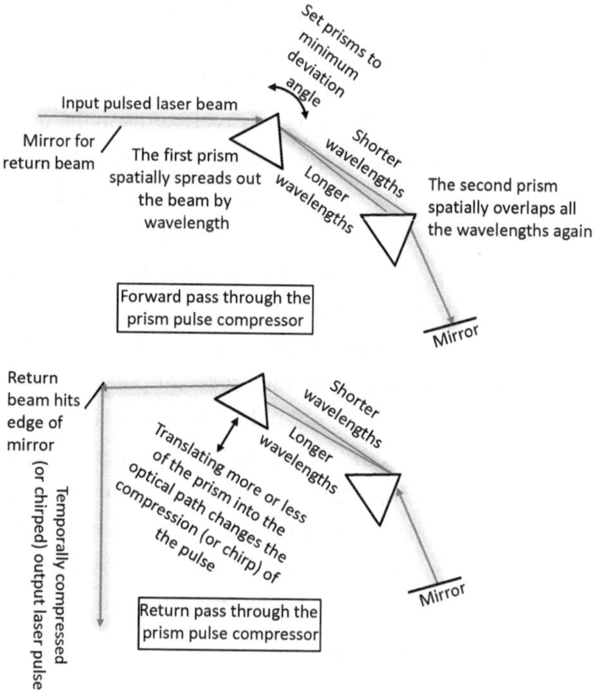

Figure 5.10

Prism pulse compressor ray diagram. Usually the pulse compressors are set up with a double-pass configuration.

beam. You will want to characterize the pulse length with an autocorellator (see Section 2.2.10 for a discussion of autocorellators). Use the feedback from the measured pulse length to optimize your pulse compressor setup.

If you don't have access to an autocorrelator, measuring the intensity of the second harmonic (Section 4.11) signal from a reference sample (such as a BBO crystal, LBO crystal, or a piece of *z*-cut quartz) will allow you to optimize your pulse compression. If you are using the second harmonic generation (SHG) approach, do be aware that SHG efficiency is polarization dependent, so think of a way to mark the orientation of the SHG reference sample and always put it in your beam path the same way.

5.8 Using a Microscope Slide as a Beam Splitter

Often it is desirable to monitor some of the wavelength or power of your excitation light in real time. Using a microscope slide will split off ~4 percent of the light and

Figure 5.11

Microscope slide used as a beam pick-off mirror or as a cheap beam combiner optic.

you can then use that light to monitor the power of your laser beam in real time and make a measurement with the other 96 percent of your light. To set up this simple beam splitter, position a microscope slide at 45 degrees to the laser beam pathway as you would a beam splitter (Section 2.3.17), and set up a detector or another optic at 90 degrees to the beam path. Beware the transparency and reflection coefficient of glass at various wavelengths. Double-check that the transmission and reflection of the glass slides you have on hand will work for your wavelength of interest.

5.9 Beam Expanders

It is often desirable to expand a laser beam's diameter to fill a lens or the back focal plane of a microscope objective. To reach a diffraction-limited spot, the laser needs to fill the whole lens. Building a beam expander is relatively straightforward. A beam expander can be built using two lenses with appropriate focal lengths to build either a Galilean or a Keplerian telescope.

The Galilean telescope uses a concave lens and a convex lens. The concave lens starts the beam diverging and then the convex lens collimates the beam again. The

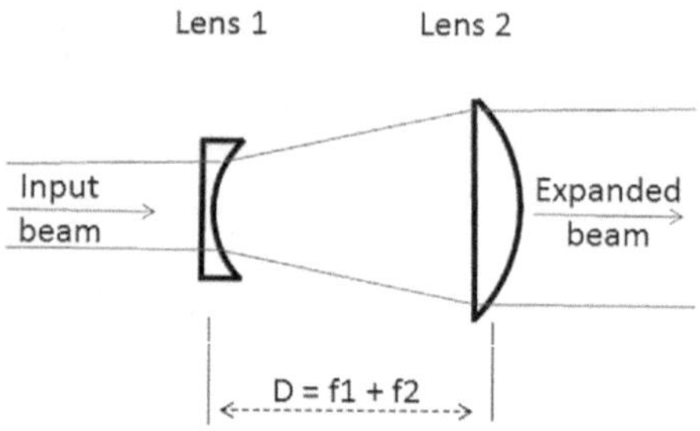

Figure 5.12

Ray diagram of a Galilean beam expander.

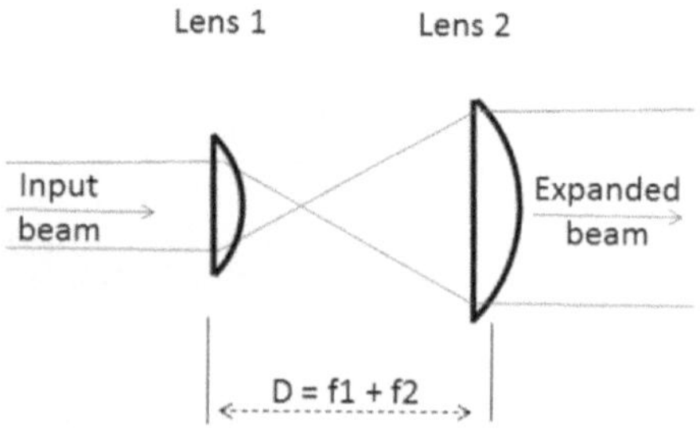

Figure 5.13

Ray diagram of a Keplerian beam expander.

expansion factor is given by the ratio of the focal lengths of the two lenses. Remember that concave lenses have a negative focal length.

$$Expansion\ factor = -\frac{f_2}{f_1} \tag{5.1}$$

The advantage of the Galilean beam expander is that the laser beam never goes through focus. For pulsed lasers, the energy density at a tight focus is sometimes enough to start ionizing the air. You will see a small white spot of light appear at the focus and usually hear a crackle of the ionized air. While this is a spectacular party trick, it is not doing you any favors in terms of laser beam quality.

The Keplerian beam expander uses two convex lenses and the laser beam goes through a focus. A pinhole positioned at the focal spot can be used as a spatial clean-up filter on the beam. The expansion factor is again given by the ratio of the focal lengths of the two lenses. Both convex lenses have positive focal lengths:

$$Expansion\ factor = \frac{f_2}{f_1} \tag{5.2}$$

Keplerian beam expanders do have the beam go through focus, so they are not usually appropriate for high power pulsed lasers. They do have an advantage for low power lasers though. You can use a pinhole as a spatial filter to clean up a messy laser beam profile by placing the pinhole around the focused spot. Any part of the laser beam that is not focusing at exactly the position of the pinhole will be rejected.

Appendices

Appendix 1

The Photoelectric Effect and Photoelectron Spectroscopy

The photoelectric effect describes how light incident on a material can cause the material to eject electrons (hence "photo" and "electric").

Electron ejection due to incident light only occurs under certain conditions. Every material has a physical property known as a work function. The work function is the amount of energy required to excite an electron from an electronic quantum state (see e.g., Section 3.3) inside the material to vacuum just outside the material.

If a material is illuminated with photons whose energy is greater than that of the material's work function, the photons can transfer their energy to electrons, allowing the electrons to overcome the work function, and leave the material.

In fact, we owe the basics found in Chapter 1 of this book to Albert Einstein. In 1905, Einstein published a paper on the photoelectric effect, for which he won the 1921 Nobel Prize. Einstein's profound insight at the time was that light's energy travels in discrete "packets" called photons that have well-defined energies. Einstein's insight came when he was trying to explain some until-then confounding experiments showing that ultraviolet (UV) light incident on metal surfaces caused the metals to emit electrons. Although the idea that light travels in discrete photons seems straightforward in retrospect, Einstein's remarkable insight opened up atomic, molecular, and quantum physics and chemistry for the likes of Neils Bohr, Louis de Broglie (see Appendices 2 and 3), Werner Heisenberg (of Uncertainty Principle fame), Paul Dirac, Erwin Schrödinger (of Schrödinger's equation, and an eponymous cat), Wolfgang Pauli (of the Pauli Exclusion Principle for electrons), Enrico Fermi (of Manhattan Project fame), etc.

In mathematical terms, Einstein wrote:

$$E_{e^-} = h\nu + \phi$$

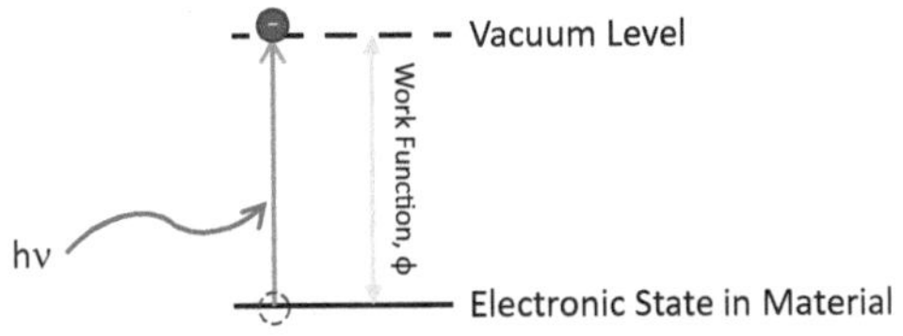

Figure A1.1

The photoelectric effect.

where E_{e^-} is the kinetic energy of the ejected electron, $h\nu$ is the energy of the incident photon (Planck's constant times the frequency of the incident light), and ϕ is the material's work function. This equality states that the energy of the emitted electron is equal to the energy of the incident photon plus the material's work function. The work function can be thought of as the potential barrier between an electron in a solid material and the electron in vacuum in front of that material. Typical work functions range from 1.5 to over 5 eV for noble metals. For example, tungsten's work function is ~4.5, whereas specialized coatings containing cesium (see below regarding photomultiplier tubes) can yield work functions as low as 1.1 eV, which is considered an extremely low potential barrier in this regime.

In reverse, the photoelectric effect can be applied to *measure* the work function of an unknown material. In photoelectron spectroscopy, a light source with known or variable energy is aimed at a material situated in a vacuum chamber. Electrons ejected from the material because of the incident light are collected by a positively biased metal electrode positioned nearby in the vacuum and measured on an ammeter as current. As the incident light energy is swept from high to low, the threshold at which no more electrons are emitted from the material corresponds to the material's work function, since at that energy or below, photons do not have enough energy to excite electrons from their energy state in the material to the vacuum. Electrons will also be ejected as the incident light energy matches the energy of various quantum states in the material; this can inform researchers' understanding of the electronic structure and/or chemical/bonding state of materials.

Photoelectron spectroscopies are typically named according to the wavelength of the excitation source: x-ray photoelectron spectroscopy (XPS), ultraviolet photoelectron spectroscopy (UPS), and visible photoelectron spectroscopy are relatively widespread, and off-the-shelf tools can be purchased, usually at an institutional level given cost (several hundred thousand dollars), complexity, and size of these machines that need to operate in ultra-high vacuum. The specifics of each of these techniques are outside the scope of this appendix, but the short version is that each can provide different information about the electronic structure and surface composition of materials to different depths below a surface and sensitivity levels depending on how the techniques are configured.

The photoelectric effect also underlies the operation of photomultiplier tubes (Section 2.3.42). As photons enter the tube, they strike a metal plate (known as a photocathode) coated with a film of material with known – and often quite low (below 2 eV) – work function. Electrons are ejected from this material, traverse a vacuum gap, and strike another metal plate, known as an electrode, which is held at a very high positive voltage bias. Because this plate is held at a high positive bias, electrons that strike this metal land at higher energy, releasing secondary electrons (Section 4.19). The yield of secondary electrons can be greater than 1, meaning that the photomultiplier tube produces more than one electron for each incident photon with energy above the work function of the first photocathode. This phenomenon is known as gain or amplification, and allows near-single photon signals to be amplified into much larger, many-electron signals. When selecting your photomultiplier tube, be sure the first photocathode has a high quantum yield in your wavelength/energy regime of interest for detection, and that you have an appropriate power supply (typically ~10–1000 volts depending on the tube specifications) for the tube. These specifications are available from photomultiplier tube suppliers such as Hamamatsu.

Appendix 2
Wave–Particle Duality

In Appendix 1 on the photoelectric effect, we discussed how light travels in discrete packets known as photons. Well, let's add another wrench into those gears. It turns out that light can be described as both discrete particles known as photons *and* using descriptions of waves.

As particles, photons have discrete, quantized energies, and via the photoelectric effect, can yield single electrons from surfaces. Nonlinear effects and electron absorption also speak to the single-packet nature of light.

However, light also diffracts from repeating structures like photonic crystals and small structures like slits (Appendix 3), suggesting that it behaves as a wave!

How do we reconcile this?

At least in this Appendix, we don't. Hopefully this is not considered a cop out by you, the reader. "Wave–particle duality" forms one of the fundamental tenets of quantum mechanics. The exact explanation has been hotly debated for over 100 years by leading physicists and philosophers, including Planck, Einstein, de Broglie, Heisenberg, Bohr, and countless others, so we will not attempt to reproduce those lengthy arguments here.

But, a key takeaway is that depending on the situation, it is convenient to describe light using wave descriptions (e.g., with wavelength) or with particles (e.g., with discrete energies) (Table A2.1). Neither interpretation is fundamentally

Table A2.1 Many optical phenomena can be described by either the particle or the wave nature of light. Some phenomena are consistent with both descriptions.

Phenomenon	Wave	Particle
Absorption	Yes	Yes
Reflection	Yes	Yes
Refraction	Yes	Yes
Interference	Yes	No
Diffraction	Yes	No
Polarization	Yes	No
Photoelectric effect	No	Yes
Blackbody radiation	No	Yes

incorrect, but both descriptions and mindsets can be leveraged to our advantage, depending on the situation.

Many scientists and engineers have a fundamental discomfort about the idea that light can be both particles and waves. We do too. But if it makes readers feel better about this, Albert Einstein also shared this discomfort, and wrote: "It seems as though we must use sometimes the one theory and sometimes the other, while at times we may use either. We are faced with a new kind of difficulty. We have two contradictory pictures of reality; separately neither of them fully explains the phenomena of light, but together they do."

Appendix 3
Young's Double Slit Experiment

Young's double slit experiment is a classic demonstration of the wave nature of light. In 1801, Thomas Young performed the first iteration of what has come to be known as Young's double slit experiment. You can perform this experiment in your lab quite simply. In this experiment, a coherent light source (e.g., a laser pointer) is aimed at a metal plate with two small slits cut in it and the light that passes through is observed on a screen placed behind the metal, opposite the light source. The slits should be placed closely together, usually within <1 mm of each other; the viewing screen should be placed several centimeters from the metal slits.

On the screen, one observes alternating intense and dim regions known as a **diffraction pattern**.

If light behaved only as particles, one would expect to see only patterns that depend on the size and shape of the slits. However, the diffraction pattern occurs because light passing through the two slits interferes with the light from the other slit. As the light propagates out of the slits at various angles, it interferes with the light leaving the other slit. Depending on the angle between the slits and the screen, the light travels a path that is equivalent to some number of light wavelengths. But, because of the small offset between the slits, the distance from the

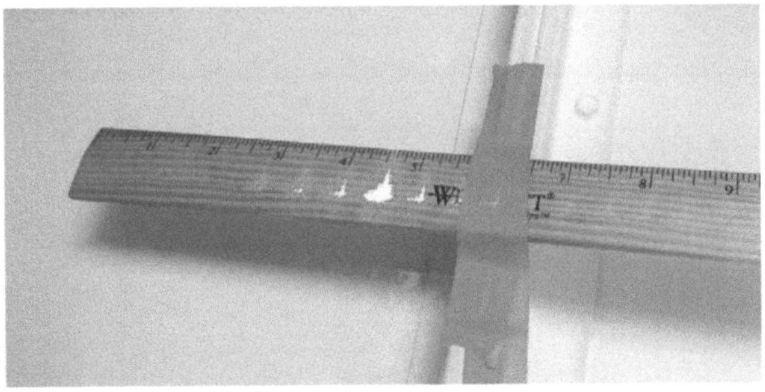

Figure A3.1

Cell phone camera image of diffracted spots on a ruler. The experiment was performed using a double slit and a 633 nm laser pointer.

first slit to the screen is slightly different from the distance from the second slit to the screen. At different points on the screen, these distances either correspond to the wave peaks of the light lining up with each other or being out of phase, which results in alternating patterns of constructive and destructive interference. This gives rise to the spots of alternating intensity on the viewing screens, and confirms the wave nature of light.

Appendix 4

Blackbody Radiation

"Blackbody radiation," also known as thermal radiation, is the light emitted from hot materials. Consider a basic example. When you run current through a fine wire, it heats up and glows red or white, as in an incandescent light bulb (Section 2.2.1). Similarly, hot coals or pokers placed into a fireplace or barbeque grill glow red or yellow. In the natural world hot lava glows red, and stars emit red, yellow, or blue light depending on their temperature.

Blackbody (or thermal) radiation is effectively the opposite process of absorption; at high temperatures (i.e., any temperature greater than absolute zero), atoms in materials are vibrating. Because atoms contain electrons, this causes the electron clouds in the materials to slosh back and forth. These "harmonic oscillators," as Planck called them, trade thermal energy for light energy, giving rise to materials glowing when they are hot. Atomic oscillations increase in frequency with greater thermal energy, so as the temperature of a material increases, the photons emitted increase in energy.

Materials only start glowing in a way that is visible to humans around ~500 °C or hotter; below this temperature, materials glow in the IR, hence using IR cameras and detectors for "night vision."

The peak wavelength at which a material glows is described by Wien's displacement law:

$$\lambda_{max} = \frac{b}{T} \tag{A4.1}$$

where λ_{max} is the peak wavelength of the light that will be emitted from a glowing body, T is the temperature in kelvins (K), and b is a constant equal to 2.8977729×10^{-3} K m. The full spectrum of light emitted from a blackbody is given by Planck's law, which can be found in physics textbooks, but contains a bit too much math to flesh out for this book.

The Stefan–Boltzmann law describes how much thermal power a blackbody emits:

$$J = \sigma T^4 \tag{A4.2}$$

where J is the thermal power emitted, σ is the Stefan–Boltzmann constant equal to 5.6703×10^{-8} W m m^{-2} K^{-4}, and T is the temperature in Kelvin. Note that the emitted thermal radiation scales with temperature to the fourth power ($\sim T^4$), so at very hot temperatures materials shed heat energy *very* readily by light emission. Your skin and clothing absorbing the IR light emitted from your hot fireplace gives rise to why you can "feel" the heat from your fireplace several feet away. Simply placing a piece of glass (which is reflective to IR light) in front of a fireplace will block the radiative heat from entering the room, hence why people use metal screens that allow light to pass through but will block flaming embers, instead of solid screens.

There is no such thing as a perfect blackbody, which is a material that emits perfectly according to Planck's law. Instead, real materials emit at a fraction of the intensity predicted at a given frequency. **Emissivity** is a concept that describes how perfect a blackbody material is. Emissivity equal to 1 corresponds to a perfect blackbody which emits at full intensity at a given frequency. Graphite, brick, soot, and ceramics have emissivity >0.9 at many frequencies, whereas polished metals have emissivity <0.1.

Index